W9-DEG-461

TEACHER EDUCATION POLICY IN THE UNITED STATES

Interest in teacher education policy is growing as educators and policymakers recognize the deep connection between excellent teaching and student achievement. What constitutes a high quality teacher education program and what standards should teacher candidates meet before receiving their teaching credential? This volume advances deep understanding of the nature and sources of policy affecting the preparation of teachers in the U.S. and the conflicts or interconnections of these policies with the broader field of education policy.

Contributions from actors in the policy world and experts representing the stakeholders are balanced and based on issues currently facing the field. Policy is viewed as evolving and political. The connection or lack thereof between policy and research is examined. Case studies ground the principles developed within specific chapters in practice and illustrate that policy questions and solutions are continually evolving and unsettled. Chapter-end commentaries by the editors relate the focus of each chapter to the overarching themes of the book: policy formation, policy influences, policy paradoxes, and connections to research.

Recent federal policy requiring highly qualified teachers in every classroom is based on a very narrow definition of teacher quality, putting pressure on state agencies to change teacher licensure and program approval policies in response to federal mandates and leading to tension between decision makers and professional educators. *Teacher Education Policy in the United States* is an essential resource for understanding and resolving today's uncertainty and confusion over teacher education policy.

Penelope M. Earley is Professor and Founding Director of the Center for Education Policy and Evaluation at George Mason University.

David G. Imig is Professor of Practice in the College of Education at the University of Maryland, College Park.

Nicholas M. Michelli is Presidential Professor at The Graduate Center of The City University of New York.

TEACHER EDUCATION POLICY IN THE UNITED STATES

Issues and Tensions in an Era of Evolving Expectations

Edited by Penelope M. Earley,
David G. Imig, and Nicholas M. Michelli

Routledge
Taylor & Francis Group

NEW YORK AND LONDON

First published 2011
by Routledge
711 Third Avenue, New York, NY 10017

Simultaneously published in the UK
by Routledge
2 Park Square, Milton Park, Abingdon, Oxon OX14 4RN

Routledge is an imprint of the Taylor & Francis Group, an informa business

© 2011 Taylor & Francis

The right of the editors to be identified as the authors of the editorial material, and of the authors for their individual chapters, has been asserted in accordance with sections 77 and 78 of the Copyright, Designs and Patents Act 1988.

All rights reserved. No part of this book may be reprinted or reproduced or utilized in any form or by any electronic, mechanical, or other means, now known or hereafter invented, including photocopying and recording, or in any information storage or retrieval system, without permission in writing from the publishers.

Trademark notice: Product or corporate names may be trademarks or registered trademarks, and are used only for identification and explanation without intent to infringe.

Library of Congress Cataloging in Publication Data
Teacher education policy in the United States : issues and tensions in an era of evolving expectations / edited by Penelope M. Earley, David G. Imig, and Nicholas M. Michelli.
    p.cm.
    1. Teachers–Training of–United States.   2. Teachers–Training of–
    Government—policy–United States.   3. Educational change–United
    States.   I. Earley, Penelope M. (Penelope McGaw), 1945–   II. Imig, David, 1939–   III. Michelli, Nicholas M., 1942-
LB1715.T4145 2011
370.71'173—dc22

                                                                    2010042964

ISBN: 978-0-415-88360-3 (hbk)
ISBN: 978-0-415-88361-0 (pbk)
ISBN: 978-0-203-84359-8 (ebk)

Typeset in Bembo
by RefineCatch Limited, Bungay, Suffolk, UK

Printed and bound in the United States of America on acid-free paper.

SUSTAINABLE FORESTRY INITIATIVE

Certified Sourcing

www.sfiprogram.org

The SFI label applies to the text stock.

On behalf of all who contributed to this volume we dedicate it to our children, grandchildren, nieces, and nephews, who are our future.

CONTENTS

PREFACE

Although much has been written about teacher education in the United States, few scholars have attempted a systematic presentation and analysis of the complex web that is teacher education policy. This volume fills that void through the presentation and analysis of a series of policy case studies, several of which recount events across a decade or more. A messy picture is revealed of multiple layers of governance and competing expectations, often colored by feelings of distrust and animosity. Initially, we anticipated the audience for this book would be graduate students studying teacher education, social policy, or teacher education policy. However, as the invited authors submitted their chapters, we became convinced that both decision makers and teacher educators should read it and engage in serious conversations about ways that policy can enable or impede the preparation of educators.

The first chapter includes a brief description of federal and state policy history, followed by a presentation of a theoretical framework we believe is useful for studying teacher education policy. Chapter 2 includes essays by Frederick Hess and Neal McCluskey written from the perspective of individuals outside of the teacher education community. Hess argues that educators should not immediately dismiss criticisms leveled at them by decision makers and others, while McCluskey recounts the expansion of the federal role in education and concludes with the assertion that, with few exceptions, federal policy has not resulted in a better education system. Five chapters present longitudinal policy case studies of events in six states: Florida, Indiana, Louisiana, New Jersey, New York, and West Virginia. There are common themes across these cases, the most jarring being policymakers' tendency to dismiss professional accountability measures and to mandate untested external ones. This point is addressed directly by Ken Zeichner's analysis of the misuse of certain teacher education assessments. A chapter by Cochran-Smith and

Fries documents a movement away from attention to social justice in education policymaking. They further argue that attention to social justice by the teacher education accreditation bodies is modest, at best. The concluding chapter, "Editors' Reflections," presents essays by each of the volume's editors on the future of teacher education in the United States.

For eight of the chapters, an editors' commentary follows the text in which perspectives on the events and themes offered by the authors are discussed and questions to promote deeper discussion of the chapters are suggested. This book is not intended to provide answers or solutions to troublesome policy conundrums; rather its purpose is to generate discussion and from that perhaps bridge the gap between the policy and teacher education worlds.

We wish to thank the authors whose work is the body of this volume. For many, the events they recounted were the cause of professional pain as they struggled to meet difficult policy mandates. Thank you does not capture our gratitude to Routledge editor Naomi Silverman. Her encouragement and wise words kept us focused and energized.

1

TEACHER EDUCATION
POLICY CONTEXT

Nicholas M. Michelli and Penelope M. Earley

Interest in teacher preparation policy has increased as educators and policy makers recognize the deep connection between excellent teaching and student achievement. Ironically, requirements of the federal No Child Left Behind Act (NCLB), which require highly qualified teachers in every classroom, adopt a very narrow definition of quality. This has resulted in tension between decision makers and professional educators over what constitutes an excellent teacher and how individuals are prepared to meet that standard. At the same time, federal laws and new expectations for federal discretionary grants have pressured state agencies to change teacher licensure and program approval policies to respond to federal mandates and challenges. This has led to uncertainty and confusion over what is a high quality teacher education program and what standards teacher candidates should meet before receiving their teaching credential. This volume explores the dynamics of teacher preparation policies proposed and enacted by the federal government and individual states. A muddled policy landscape is revealed in which laws and regulations may be inconsistent; are rarely informed by empirical evidence; are unable to keep up with changing demands by decision makers; and by accident or design may ultimately harm vulnerable school populations.

In this chapter, we briefly consider late 20th-Century teacher preparation policy history in the United States, looking first at selected federal actions and then considering illustrative state teacher education policies. Following that discussion, we offer a theoretical framework to analyze contemporary teacher education policy issues. One theme of note in federal and state policy is that teacher education policy tends to either be disconnected from or to lag behind other education policy initiatives even though educators are expected to immediately implement significant policy changes imposed on K-12 school systems. As a consequence of these disconnects and lags, teacher preparation programs have been criticized for being out of touch or unresponsive when in reality the dilemma was and is one of how to catch up. This has led critics to recommend abandoning college and university-based teacher education in favor of an open market approach in which new teachers would receive expedited, shorter preparation and be able to enter the classroom sooner. Although a number of these expedited programs have been established the jury is out in terms of evidence that they are more nimble in responding to federal, state, or local policy changes affecting K-12 classrooms (Sykes & Dibner, 2009).

Seven of the eight of the chapters that follow are written by individuals engaged in the everyday work of teacher preparation. They reflect recurring and new policy dilemmas and challenges. For each chapter, the editors offer a brief commentary designed to illustrate cross-cutting themes and connections and to challenge readers to consider policy alternatives. A concluding chapter discusses themes and policy paradoxes that emerge from these descriptions of policy events and raises questions we hope will advance policy deliberations.

Teacher Preparation Policy History and National Context

Following World War II, as a result of the Serviceman's Readjustment Act of 1944 (G.I. Bill), the number of veterans seeking a postsecondary education increased. Lawmakers initially predicted that fewer than 10% of servicemen would apply for G.I. Bill benefits; however, by 1950 over two million veterans were enrolled in some form of postsecondary education (Thelin, 2004, p. 263). The influence of the G.I. Bill on teacher preparation was manifest in several ways. In the first part of the 20th Century, some teachers attended universities and received bachelor's degrees; however, many were educated in institutions known as teachers' colleges or normal schools. Initially, teachers' colleges offered certificates rather than degrees, but as demand for college options increased these institutions expanded their programs to become bachelor's degree granting state colleges. The transformation of teachers colleges to four-year state colleges was accelerated by the arrival of the returning servicemen who expected to attend college, but not necessarily to become teachers. As Lucas (1997) notes, teachers' colleges quickly refashioned themselves into multipurpose state colleges. Furthermore, with the creation of comprehensive state colleges, teacher education programs found themselves part of a larger institutional enterprise and needing to fight with other units for resources and academic legitimacy; a tension that continues today.

The postwar baby boom carried with it the expectation of parents who attended college with assistance from the G.I. Bill that their children would follow a similar path. This not only led to the need for more elementary and secondary school buildings but also created an assumption that these schools would prepare more students to pursue education beyond high school. As a result, the demand for more teachers who would be able to prepare these students for admission to college increased. Thus, the G.I. Bill initiated a slight shift in the purpose of K–12 education. It was not just to provide the computational and literacy skills needed for high school graduates to be employable, but also to prepare more of them for postsecondary education.

The 1950s saw an expansion of the federal role in education with the 1958 enactment of the National Defense Education Act (NDEA), which among other provisions encouraged persons to enter teaching in mathematics, science, and foreign languages (Cross, 2004; Earley & Schneider, 1996). This multifaceted law was based on the premise that the nation's security was at risk, in part because of the launch of the Soviet satellite *Sputnik* the previous year, and that it was in the national interest for the federal government to leverage education policy. In the teacher education realm, this was translated into funding for summer teacher training institutes in mathematics and science, many offered through the National Science Foundation (NSF) or molded after them. As a policy tool, we offer three observations about the teacher development programs in NDEA. First, professional development was not made available to all teachers, only a limited number of them. Second, political support for NDEA was predicated on the assumption that the Soviet Union's education system was vastly superior to that in the United States—a questionable assumption. And, third, an evaluation by the General Accounting Office (now renamed the General Accountability Office) found that the NSF or NSF modeled teacher institutes had no effect on student learning in mathematics and science (GAO, 1984). NDEA is an example of the challenge of policy lag. That is, lawmakers wanted to quickly shore up national security in part through the development of an aggressive space exploration program. NDEA provided professional development for a small number of science teachers in the hope that the children they taught would become interested in and adept in mathematics and science. Even if there were strong evidence that the NDEA teacher institutes had broad impact, many years would pass before the children being taught would become rocket scientists.

During the 1960s and 1970s the federal government's direct role in teacher education was limited to support for some small categorical programs—such as Teacher Corps. Teacher Corps was legislated in 1965 as a provision in the Higher Education Act to support partnerships between institutions of higher education and school districts to provide teacher education coursework and experiences for idealistic young men and women to work in high poverty schools. Over the 15 years that Teacher Corps was supported as a federal program its provisions and focus shifted as lawmakers' expectations of it changed. It is important to note that Teacher Corps was a small program, with fewer than 10% of institutions and school districts funded by it (Jordan & Borkow, 1985). Evaluations of Teacher Corps found that it did recruit individuals into teaching and many of those individuals went into difficult to staff schools, but it is not clear whether or not the people who participated in the Teacher Corps program would have become teachers in any event. As Sykes and Dibner (2009) report, the expectation that Teacher Corps graduates would be able to "change the schools and the profession proved unrealistic" (p. 16). Moreover, many of the partnerships between higher education and K-12 schools created with Teacher Corps grants ended when federal funding stopped.

Another program to support teacher preparation was the Trainers of Teacher Trainers Program, created as part of the Education Professions Development Act of 1967. Like Teacher Corps, this was a modest categorical federal program that awarded grants, in this case to colleges and universities, to build community/school/institutional collaborations to prepare teachers (Sykes & Dibner, 2009). The Trainers of Teacher Trainers Program was discontinued in 1973 and, according to Sykes and Dibner, evaluations found no impact as a result of the grants.

The NDEA summer teacher training institutes, the Teacher Corps Program, and the Training of Teacher Trainers Program were not the only federal interventions in the 1960s and 1970s to focus on teacher education but they were the largest and are notable as illustrations of the disconnect and lag between national education policy and teacher education policy. Supporters of NDEA expected a rapid turn around in mathematics and science skills in American children. But by the time the NDEA summer institutes were in place, and teachers received professional development and modified their instructional skills, the U.S. had already launched its own space program. In addition, the summer institutes only reached a small number of the nation's teachers. Teacher Corps was enacted at a time when Congress and then President Lyndon Johnson envisioned a federal government that could create and sustain what was known as the Great Society. Teacher Corps and to some extent the short-lived Training of Teacher Trainers were expected to generate teachers for difficult-to-staff urban and rural schools who also would be community change agents. However, neither program had the capacity to have a measurable nationwide impact on the problems facing low income and urban communities and schools.

The National Defense Education Act clearly set out congressional intent that a strong education system is a matter of national security and as such attention to K-12 education is a federal priority. Thus, after the late 1950s federal lawmakers and others kept a watchful eye on the quality of schooling in the United States, and they did not like what they saw. The strongest criticism of American education came in the form of *A Nation at Risk: The Imperative for Educational Reform* (National Commission on Excellence in Education, 1983). Using sound bite rhetoric such as "If an unfriendly foreign power had attempted to impose on America the mediocre educational performance that exists today, we might well have viewed it as an act of war" (p. 5), writers of the report leveled sharp criticisms at schools, the curriculum, and educators. They went on to argue that a strong education system is necessary for the United States to compete in global markets. With this language the national purpose of education became not just one of promoting national security but of promoting economic security as well.

A Nation at Risk gave voice to concerns about the K-12 education system but it was not the only vehicle to raise doubts about education quality. Decision makers pointed to increased federal support for K-12 education—in particular, through laws such as the Elementary and Secondary Education Act (ESEA) and the Individuals with Disabilities Education Act (IDEA) – and the growth and expansion of federal, state, and local bureaucracies to oversee these investments. With all of this apparent oversight and increases in government support for education, critics asked: Why was there such a gap in achievement between children in wealthy and poor communities? Why did more white children complete high school than black or brown children? Why did children in the United States score lower than other nations on international examinations? By focusing these questions on the education community, the critics "reinforced the idea that social and economic causes of poverty can be discounted as causes of poor performance" (Sunderman, 2009, p. 12).

An attempt began to address these questions in 1989 with what became known as the Charlottesville Education Summit, during which President G.W. Bush and the nation's governors met to agree on common policy goals to improve the nation's schools. Initially, they adopted six national goals (ultimately two more were added), and in 1990 the National Goals Panel was established to monitor success in meeting these goals (Cross, 2004; Vinovskis, 2009). One of the eight national goals addressed teacher quality, stating in part "our nation's teaching force will have access to programs for the continued improvement of their professional skills needed to instruct and prepare all American students for the next century" (Vinovskis, 2009, p. 71). With the addition of this goal one might expect that teacher education policy would be carefully aligned with other education policy initiatives. That was not the case. In the decade following the Charlottesville Summit a variety of federal programs were authorized and funded to improve the education of children with disabilities, target schools with large concentrations of poor children, provide support to help children learn English, and strengthen mathematics and science achievement. But with few exceptions, these policies assumed that the teacher preparation system would automatically and quickly change programs so all teachers would be able to respond immediately to each new federal mandate. As many of the cases presented in the chapters that follow illustrate, it is difficult for teacher education, wherever it occurs, to respond immediately to policy expectations.

It may be argued that no teacher education system could ever be nimble enough to quickly fulfill expectations of the policy community. Teacher educators would further assert that a system that can change on a dime in response to policy shifts from state or federal decision makers is inconsistent with educator preparation programs based on agreed upon professional standards. This line of reasoning did not sway critics of K-12 schools and of teacher education. In 1996 The National Commission on Teaching and America's Future (NCTAF) put forward a vision of reform for teacher education based on a strong framework of accreditation and teacher licensure with the aim to truly professionalize teacher education. Critics were quick to agree with the problems identified by NCTAF but dismissed the proposed remedies as reinforcing those aspects of the system—teacher licensure and preparation program accreditation—that they felt militated against real reform (Ballou & Podgursky, 1997). The solution the critics suggested was to deregulate the teacher preparation and licensure systems and let school districts make hiring decisions not constrained by requirements that teachers complete an approved or accredited teacher education program and hold a state-issued license (Hess, 2001). As Sykes and Dibner (2009) suggest, this debate became, and continues to be, one between supporters of professional or market based agendas.

Pointed discussions about how best to recruit and prepare teachers were quite public and did not escape notice by federal policymakers. As a result, the 1998 reauthorization of the federal Higher Education Act (HEA) and 2001 passage of amendments to the Elementary and Secondary Education Act creating the No Child Left Behind Act established a federal presence in state policy to prepare, license, and evaluate the quality of teachers. This was done in the Higher Education Act through provisions in Title II of that law requiring that colleges and universities send the state the pass rate of their students on the state's teaching license examination. The state was required to make this information available to the public and to forward it to the federal government. The thinking behind this part of the law was based on the idea of choice in the marketplace. Policymakers reasoned that individuals who want to be teachers would use pass rate information to decide where to enroll. The flaw in the logic behind this mandate is the assumption that the best teachers are those from institutions with high pass rates and the weakest from institutions with low pass rates. Some lawmakers mistakenly believed that the law would prevent students who did not pass the licensing examination from becoming teachers. However, no individual can become a teacher without first passing the state exam, so the thinking that this provision alone would keep unqualified persons from entering teaching was misguided.

Four years later, in 2002, when NCLB was enacted, states receiving funds through it were expected to assure that all teachers met the federal definition of highly qualified included in the law. Central to the highly qualified definition was that the teacher should either have an undergraduate major or minor in the subject to be taught or be able to demonstrate subject matter competency through an examination or other measures. This provision was in response to data suggesting that many educators were teaching out of field (Ingersoll, 1999). That is, an individual with a degree in English might be pressed to also teach world history, a situation common in small schools where enrollments are low. The new federal definition of teacher quality, coupled with an NCLB mandate that all children would demonstrate proficiency on their state's academic examination within 12 years, put added pressure on educators and the programs that prepared them.

By the first decade of the 21st Century, more policy attention was being directed to teacher education by the federal government. Because constitutionally states have primary authority over education in the United States, federal implementation of teacher education policy was indirect. This was done by linking substantial federal funds available for states or to institutions of higher education to changes in state policies. In the Higher Education Act, colleges and universities that receive any federal funds were required to gather and report pass rates on the license examination for their teacher education students or risk loss of millions of dollars. In the No Child Left Behind Act, federal money sent to states to help children in poverty could be withheld if the state did not revise teacher licensing regulations to be consistent with the highly qualified teacher definitions. In essence, within a five-year period, policies to govern teacher education were firmly entrenched in federal laws. In the chapters that follow, the impact of these federal policies on the day-to-day work of preparing teachers is described, and a paradox of what is meant by teacher quality and how to measure it emerges.

State Policy Context

As we have noted, at the beginning of the United States of America, education was a function of the states and not the federal government. Nowhere in the Constitution of the United States of America does the word "education" appear. According to the 10th Amendment to the Constitution functions not delegated to the federal government nor prohibited to the states go to the states. Thus, most state constitutions provide for the oversight of the K-12 public education system through the establishment of state boards and agencies. In general, authority to award teaching licenses and review teacher education programs is not detailed in state constitutions but over time has been codified in state statutes. Before the point when state agencies developed requirements for teachers, standards and expectations were set by the local schools, the city, or the county. Consequently, the regulation of teacher preparation and the standards for receiving a teaching credential have a long history of state and local governance.

Before the ratification of the U.S. Constitution most colonies gave responsibility to ministers, elders, citizens, and lay boards to find and hire teachers. Criteria used by these private reviewers included courage, strength, their beliefs and values, and sometimes their academic knowledge. As Sedlak (2008) observed, often close friends and relatives were hired for these positions. It was not until the second half of the 19th Century that teacher hiring policy, and hence teacher education, moved to counties primarily and occasionally to the state level. Yet, even at that time, teaching often was considered a patronage position, with jobs awarded by local officials on the basis of relationships rather than qualifications. During that time normal schools emerged—teacher education institutions that provided a high school education for teachers (Wilson & Youngs, 2005). State control of the certification process grew dramatically between 1898 and 1937. The shift from local to state control is summarized in Table 1.1 (Angus & Mirel, 2001).

TABLE 1.1 Type and Number of State Systems of Teacher Certification, 1898–1937

	1898	1911	1921	1926	1937
State systems (state issues all certificates)	3	15	26	36	41
State-controlled systems (state sets rules, conducts exams, county issues some certificates)	1	2	7	4	3
Semi-state systems (state sets rules, writes questions, county grades papers, issues certificates)	17	18	10	5	1
State-county systems (both issue certificates, county controls some certificates)	18	7	3	2	2
State-local system (full control by town committees)	2	2	2	2	1

Thus, by 1937, 44 states had state controlled or state dominated systems (Angus & Mirel, 2001). During this early period tests were sometimes required, sometimes not required, with the recommendation of an approved institution of higher education becoming the basis for issuing a certificate. In fact, by 1937, 28 states which had teacher tests had abolished them (Wilson & Youngs, 2005).

Across the states most policy affecting teacher education is established by the state departments of education (DOEs), staffed by appointees rather than elected individuals, and with the final decisions on policy resting primarily with state boards of education. Currently state board members are appointed in 35 states with members elected in 13 and in two states the boards are a combination of elected and appointed members. There has been discussion of the fact that in many states policy comes not from elected officials but rather from appointed members of the executive branch. Across the states, however, a system of regulation notification has evolved with departments of education publishing notifications of rule changes prior to final approval, and often publishing the comments received before final action. Thus, there may be an avenue for public input but it differs from the manner in which voters hold elected officials accountable. The power of regulatory action is illustrated in the policy case studies of Louisiana, Florida, and Indiana in Chapters 3, 4, and 6.

In every state the authority for teacher education policy now is found at the state level, either in the hands of a state board of education through a state department of education or an independent state standards board.[1] State standards and practices boards emerged in the 1980s in part as an effort to move teaching to the status of a profession with the view that most professions had control of their own regulation. The hope on the part of teacher groups, including the National Education Association, was to foster independent boards that would make decisions about certification requirements and standards independent of the state board of education. Those remaining include California, Delaware, Georgia, Hawaii, Iowa, Kentucky, Minnesota, North Dakota, Oregon, and Wyoming. Semi-independent boards are found in Nevada, North Carolina, Delaware, Maryland, and Texas. Interestingly the most recent data on the identification of standards boards in the United States comes from a report that studied the state auditors' recommendation to eliminate the Hawaii independent board (Chung and Kodama, 2010).

As discussed earlier, for the most part there was little federal involvement in the regulation of teacher education, with occasional funding to attract teachers into needed areas of specialization including the National Defense Education Act and the Teachers Corps program. A major break from this distribution of power came with the reauthorization of the Higher Education Act in 1998 (HEA). As part of the reauthorization, Title II of the law required colleges and universities to report pass rates on content knowledge tests for teacher education candidates, along with other data, to the state, where it would be compiled and sent to the U.S. Department of Education. The Department of Education was to use these data for an annual report to Congress on the status of teacher education. This provision in HEA, Title II, requiring public disclosure of teacher candidates' pass rates on the state licensing examination, brought teacher education to the attention of college presidents perhaps as never before. Congress' expectation was that there would be comparisons of the success of colleges on this measure and public disclosure of pass rates to prospective students and the public. The anticipated effect of such comparisons never materialized. There was no standard teaching cut-off score, so a pass rate in one state could not be directly compared with a pass rate in another state and furthermore there were different tests in different states, making comparison across states impossible. This did, however, mark the beginning of more serious federal involvement in policy affecting teacher education and this was done by linking the federal funding to colleges and universities to compliance with requirements to submit reports on their teacher education programs to state officials. There were other elements of the reauthorization of the Higher Education Act that affected teacher education as well, including a focus on alternate routes to certification with some pressure to allow organizations other than colleges and universities to offer teacher preparation programs.

When Congress reauthorized the Elementary and Secondary Education Act and renamed it the No Child Left Behind Act (NCLB), a definition of a highly qualified teacher was included in Title I. States were required to assure in policy that all teachers met the federal highly qualified definition if the state expected to receive NCLB funds. The definition was vague and essentially required that a teacher hold a certificate approved by the state which would have included passing a content test. Nevertheless, there were serious impacts in some states, especially in many rural states and states with large urban populations. In New York City, for example, until the passage of No Child Left Behind it was common practice to employ uncertified teachers. With the end of that option, alternative routes to certification flourished, since NCLB considered an alternative certificate issued by the state as evidence of high quality (Michelli, 2005).

When the second Bush administration ended and the Obama administration began, the reauthorization of NCLB was already overdue. Educators and policy analysts predicted that many provisions in NCLB would change. As it turned out. the Obama approach to policy was revealed through several major federal grant programs, the largest of which, Race to the Top, was aimed at states. Grant applications could earn points in a number of categories, but one absolute requirement was the absence of any statutory or other barrier to using value-added outcomes to judge the quality of educators in personnel actions. Chapter 3 describes one state's attempt to use value-added assessments to differentiate successful teachers from unsuccessful ones. In Chapter 5, Zeichner questions the use of these measures absent empirical evidence that they actually are able to link teacher actions to student outcomes.

The economic downturn in 2008 that affected many state budgets pushed nearly every state to apply for funding through Race to the Top and also led to hurried efforts to change policy either by state board or legislative action. Among the areas of focus with potential impact on policy in teacher education were removing barriers to value-added assessment of teachers and candidates, lifting or raising the caps on charter schools, developing longitudinal data systems to allow for the comparisons across colleges and school systems, and changing policy to allow organizations other than colleges and universities to prepare teachers. In Chapter 3, on Louisiana, the effects of a value-added comparison system in one state are explored. Zeichner's chapter (Chapter 5) also addresses the use of value-added approaches to evaluate teachers.

Policy Framework

Even though decision makers have become more aggressive in legislating teacher qualifications and preparation requirements, the scholarship on teacher education policy has been limited. Political scientists are concerned with developing models or theories to explain what happens in the policy world (Heck, 2004; Sabatier, 2007), but the dispersed nature of teacher education policy, the fact that policies in one area may contradict policies in another, the problem of policy lag, and the status of teacher education in the United States suggest that for the most part political science models do little to help explain what happened or suggest what might happen in the policy future. Datnow and Park (2009) offer three theoretical frames to study contemporary education policy issues: the technical-rational model, the mutual adaptation model, and the sense-making/co-construction model. Of these three the sense-making/ co-construction model has the potential to help illuminate implementation of teacher preparation policies; they write, "[s]ense-making theories underscore the complex interrelationship between meaning and action" (pp. 350–351). It anticipates that the policy researcher will attend closely to perceptions and interpretations of events by those involved, and as such it allows the analyst to account for values and beliefs of those engaged in policymaking. Co-construction adds to this attention to political and cultural issues and power relationships (p. 351). To this sense-making co-construction theory we add the work of Deborah Stone (1997), who suggests that the values undergirding the American political system are, in reality, paradoxical. As an example, American citizens believe the ability to vote for those who will lead the nation is a cherished right. However, citizens also value a Constitutional provision—the Electoral College—that may lead to a presidential election won by the individual with fewer popular votes. Stone further argues that the tools used to attempt to manage this paradoxical system of American governance are complex social processes that often cannot be understood by rational analysis alone. Together sense-making theory and policy paradoxes create a useful framework for thinking about teacher education. Specifically, by employing the sense-making approach, is it possible to uncover policy paradoxes and better understand the dynamics of teacher education policies?

As noted previously, teacher education is governed by many masters. At the institutional level, there are policies requiring data gathering and meeting curricular standards for purposes of accreditation, a non-governmental process. At the same time, there are parallel but not identical accountability requirements mandated by state and federal governments. Clearly, the state remains an important player because standards for allowing a college or university to offer an approved teacher education program are in state law or regulation. In addition, teacher education programs must ensure that their programs include the professional and content knowledge outlined in state law for students to pass appropriate examinations and receive a teaching license. Traditionally the federal government's influence on teacher education has been indirect by offering categorical grants or contracts for teacher education programs or faculty to do certain things, such as establish partnerships with K-12 schools or conduct research on ways to prepare teachers to teach English language learners. However, in the past decade the federal government has assumed a more aggressive role in teacher education policy by mandating extensive data about teacher education graduates—called program completers in federal higher education law—as well as through the highly qualified teacher provisions in NCLB.

In the chapters that follow, authors discuss the shifting federal role, the nexus of teacher education policy with social justice, the tenuous link between empirical research and policy, and describe teacher education policy cases. These are not case studies in the traditional sense because the authors were not asked to employ case study methodologies. Rather authors were invited to detail a policy event or events related to teacher preparation in which each was a participant or close observer. Three of these case studies span a decade or more and as such respond to the call for longitudinal policy case studies (Cohen-Vogel & McLendon, 2009). Following each chapter is a brief commentary, written by the editors, in which we apply the theoretical frameworks of sense-making and policy paradoxes to analyze the case. Together the policy case study and commentary are intended to generate discussion of policy decisions and alternatives by decision makers, scholars, and students of teacher education.

In the second chapter, two essays—one by Frederick Hess and the other by Neal McCluskey—discuss and document the expanded authority of the federal government over education. Chapter 3 documents policy events in Louisiana that ultimately led to adoption of a Value Added Teacher Preparation Assessment. Changing policy directions emerge in the Florida study presented in Chapter 4. In that chapter, authors detail the often rapidly shifting policy tides and how efforts to increase the supply of new teachers created new institutional challenges. Following discussions of events in Louisiana and Florida, in Chapter 5 Ken Zeichner offers a perspective on teacher assessment policy based on experiences in Wisconsin and Washington. He concludes that absent an empirical base for policies to measure teacher effectiveness, the impact of these policies is open to question. Zeichner suggests that forging closer ties between teacher preparation institutions and K-12 teachers is essential. Chapter 6, an analysis of the creation and demise of an independent professional standards and licensing board in Indiana, and Chapter 7, a description of a K-16 partnership in New Jersey, raise the question of why some partnerships appear to be sustainable whereas others are not. Authors Dempsey and Shanley also describe K-12 partnership efforts (in Chapter 8), and details the different ways policy plays out in urban and in rural contexts. In Chapter 9, Cochran-Smith and Fries draw us back to the purpose of education in the United States through an exploration of how matters of social justice are included or excluded in teacher education policy. The concluding chapter is three essays, each written by one of the volume's editors, adding concluding observations on the chapters and how events described in the chapters may be explained by or inform policy theory. A glossary of terms that are important to understanding teacher education policy is included after Chapter 10.

Note

1 In the NEA's classification there are a series of conditions that must be met for a board to be independent, ranging from authority over standards, budget, fees, the authority to issue, renew, and revoke licenses and the authority to approve teacher education programs. Those listed as semi-independent lacked one of more of these elements of authority (NEA, 2003).

References

Angus, D.L. & Mirel, J. (2001). *Professionalism and the public good: A brief history of teacher certification.* Dayton: The Fordham Foundation, p. 16.

Ballou, D. & Podgursky, M. (1997). Reforming teacher training and recruitment: A critical appraisal of the recommendations of the national commission on teaching and America's future. *Government Union Review* 17(4), 1–53.

Chung, C. & Kodama, K. (2010), The Hawaii Teacher Standards Board: Is oversight needed? Honolulu: Legislative Reference Bureau. Retrieved September 5, 2010 from http://lrbhawaii.info/reports/legrpts/lrb/2010/hiteachbrd.pdf

Cohen-Vogel, L. & McLendon, M. (2009). New approaches to understanding federal involvement in education. In G. Sykes, B. Schneider, & D.N. Plank (Eds.), *Handbook of education policy research* (pp. 735–748). New York: Routledge.

Cross, C. (2004). *Political education: National policy comes of age*. New York: Teachers College Press.

Datnow, A. & Park, V. (2009). Conceptualizing policy implementation. In G. Sykes, B. Schneider, & D.N. Plank (Eds.), *Handbook of Education Policy Research* (pp. 348–361). New York: Routledge.

Earley, P. & Schneider, E. (1996). Federal role in teacher education. In J. Sikula (Ed.), *Handbook of research on teacher education*, 2nd edition (pp. 306–319). New York: Macmillan.

General Accounting Office (1984, March 6). *New directions for federal programs to aid mathematics and science teaching* (GAO/HEHS-94–208). Washington, DC: Author.

Heck, R. (2004). *Studying educational and social policy*. Mahwah, NJ: Lawrence Erlbaum Associates.

Hess, F. (2001). *Tear down this wall: The case for a radical overhaul of teacher certification*. Washington, DC: Progressive Policy Institute.

Ingersoll, R. (1999). The problem of underqualified teachers in American secondary schools. *Education Researcher 28*(2), 26–37.

Jordan, F. & Borkow, N. (1985). *Federal efforts to improve America's teaching force* (CRS Report No. 85–644 S). Washington, DC: Congressional Research Service.

Lucas, C. (1997). *Teacher Education in America*. New York: St. Martin's Press.

Michelli, N. (2005). The politics of teacher education: Lessons from New York City. *Journal of Teacher Education 56*(3), 235–241.

National Commission on Excellence in Education (1983). *A nation at risk: The imperative for educational reform*. Washington, DC: Author.

National Commission on Teaching and America's Future (1996). *What matters most: Teaching for America's future*. New York: Author.

National Education Association (2003). *Report on the status of professional boards of teaching in the United States*. Washington, DC: Author.

Sabatier, P. (2007). *Theories of the political process*, 2nd edition. Boulder, CO: Westview Press.

Sedlak, M.W. (2008). Competing visions of purpose, practice and policy: The history of teacher certification and the United States. In M. Cochran-Smith (Ed.), *Handbook of research on teacher education* (p. 587). New York: Routledge.

Stone, D. (1997). *Policy paradox: The art of political decision making*. New York: W.W. Norton & Co.

Sunderman, G. (2009, Summer). The federal role in education: From the Reagan to the Obama administration. *Voices in Urban Education 24*, 6–14.

Sykes, G. & Dibner, K. (2009). *Fifty years of federal teacher policy: An appraisal*. Washington, DC: Center on Education Policy.

Sykes, G., Schneider, B., & Plank, D.N. (Eds.) (2009). *Handbook of research on education policy*. New York: Routledge.

Thelin, J. (2004). *A history of American higher education*. Baltimore, MD: Johns Hopkins University Press.

Vinovskis, M. (2009). *From a nation at risk to no child left behind*. New York: Teachers College Press.

Wilson, S. & Youngs, P. (2005). Research on accountability processes in teacher education. In M. Cochran-Smith & K. Zeichner (Eds), *Studying teacher education* (p. 594). New York: Erlbaum and AERA.

2

PERSPECTIVES ON FEDERAL POLICY

FOOT-DRAGGING AND OVERREACHING

The Story of Federal Teacher Policy

Frederick Hess

In 2010, the teacher policy landscape was marked by several dramatic developments related to value-added assessment and the Race to the Top. Colorado passed state senator Mike Johnston's much-discussed Senate Bill 10–191, which overhauled the state tenure system and required all districts to make use of student learning when evaluating teacher performance. In response to the federal Race to the Top program, a number of states removed data "firewalls" that had prohibited educators from linking student achievement to individual teachers.

But the two most notable developments were probably the near-passage of Florida's Senate Bill 6 and the decision of the *Los Angeles Times* to calculate and then run value-added data for individual teachers. In Florida, the Republican legislature passed a bill that would have abolished tenure, mandated the creation of assessments that could calculate value-added scores for all of the state's teachers, and then required that teacher pay and retention decisions be driven by student performance on these assessments. The only thing that prevented the bill from becoming law was the decision of embattled Republican governor Charlie Crist to veto it, as a prelude to his decision to run for Florida's U.S. Senate seat as an independent with the backing of the Florida Education Association.

When it came to the *L.A. Times*, the paper used seven years of reading and math scores from the L.A. Unified School District (LAUSD) to calculate performance for individual teachers who'd taught Grades 3 through 5. The research touched off a national furor. In particular, experts noted the technical challenges with calculating individual-level value-added in that fashion and the structural problems (such as those posed by students who receive substantial pull-out instruction or work with a designated reading instructor). Such structural questions mean that teachers who are producing substantial gains might be pulled down by inept colleagues, or that teachers who are not producing gains might look better than they should.

In both cases, the smart, disciplined use of value-added systems to evaluate teachers were overtaken by a crusading temptation to wield new tools with too heavy a hand. These efforts put more stress on primitive systems than they could bear while promising to unnecessarily entangle a useful management tool in personalities and public reputations. Unfortunately, these little episodes are par for the course when it comes to teacher policy.

For more than two decades, we have seen efforts to rethink the way K-12 handles staffing quickly morph from useful innovations into caricature. Sensible ideas are too often met with howls of protest and determined pushback, frustrating would-be reformers and leading them to finally disregard more sensible concerns from educators while embracing over-the-top, ham-handed policies as the only opportunity to upend the status quo. This is more or less the story of the past quarter-century when it comes to federal efforts to improve teacher quality at the federal level.

Today's debate about teacher quality may seem new to policymakers or reformers, but it is part of a much longer narrative. In the late 19th and early 20th Century, a loose-knit national network of professors, administrators, and state officials sought to standardize existing, erratic arrangements. Between about 1890 and 1940, these "professionalizers" succeeded in formalizing licensure at the state level, while increasingly linking licensure to the completion of "accredited" preparation programs predominantly staffed by professionalizers and sympathetic faculty.

After World War II, the National Education Association and allied organizations established the National Commission on Teacher Education and Professional Standards (TEPS). In 1952, TEPS, along with the American Association of Colleges for Teacher Education and the National Association of State Directors of Education, founded the National Council for Accreditation of Teacher Education (NCATE) to accredit teacher preparation programs. For decades, these same organizations have represented the institutional voices of teacher educators and teacher training. Their notion of professional responsibility and improved quality naturally favored increased support for and deference to teacher education programs and state certification. They favored accreditation and new federal dollars for teacher education, but resisted competitive grants or proposals that programs should be required to report outcome metrics.

This was the established approach to the teaching challenge, and it made good sense given an abundance of aspiring teachers. As the labor market started to shift in the post-World War II era, however, and as routine assessments of student learning became a feature of the education landscape in the 1970s, the reliance on these traditional models and input metrics came to seem more problematic.

During the 1980s, *A Nation at Risk*'s indictment of teacher quality gave rise to heated debates about how to attract and retain good teachers. One response to this challenge was to seek to specify new guidelines that would toughen up the existing licensure and preparation system. This was the tack of influential reports by the Carnegie Task Force and the Holmes Group, which called for more required training, more funding for teacher preparation and teacher salaries, more integrated subject-matter training, higher standards, and a tiered career ladder. Born of these efforts was the National Council for Teaching and America's Future (NCTAF), which soon became the self-professed champion of "professionalization."

From its inception, however, NCTAF's push for more demanding standards and heightened quality also represented an effort to shutter or stifle the alternative teacher licensure programs that had started to emerge in the 1980s—most notably in New Jersey. In the 1990s, NCTAF generally opposed the fuzzy-faced Teach for America and proposals to make it easier to remove teachers or to link teacher compensation to student performance.

NCTAF's "professionalization" agenda proved politically useful for the established teaching community by providing a common, professionally endorsed, widely acceptable platform for governors, university presidents, and education school deans that called for channeling more resources to teacher preparation and boosting the field's prestige. The recommendations, however, did little to assuage critics concerned about teacher preparation, the culture of teaching, that the profession had not evolved in response to a changing labor force, or that it failed to address professional incentives or accountability. Moreover, the focus on traditional gatekeepers and calls for increasing preparation program requirements actually aggravated the concerns voiced by critics.

Those critics would push back with increasing vehemence in the years that followed. Viewing the efforts of professionalizers as a defense of the status quo, they instead started to suggest holding teacher education programs accountable for the quality of their graduates, that licensure barriers be lowered or eliminated, that teacher evaluation and pay be based on student learning, and that teacher tenure be diluted or ended.

During the 1994 reauthorization of ESEA, particularly in the companion Improving America's Schools Act, the Clinton administration offered dollars for states to demonstrate that Title I funds were being put to good use by encouraging states to adopt voluntary report cards on student performance. States were to issue toothless reports on student performance by regularly assessing students at least once during the elementary, middle, and high school grades. The response from the professional education community was furious pushback, denouncing this desire for testing, declaring it a poor and unreliable representation of what schools and teachers were doing, and thus infuriating a wide swath of policymakers.

By 2001, when the Bush administration's No Child Left Behind proposal framed the reauthorization discussion, progressives like Senator Ted Kennedy and

Representative George Miller (ranking Democrats in the Senate and House education committees, respectively) and Republicans in the Bush administration and on Capitol Hill were ready to discount the doubts and complaints of teacher unions, education schools, and professional educators. Trust in the expertise and intentions of these groups was thin, at best. The result was that the players had little use for even sensible "establishment" critics of the NCLB testing and accountability scheme, yielding a law that was long on fervor and good intentions but short on appreciation for what its accountability mechanisms could and couldn't do.

Disputes about how to improve teaching achieved a new national prominence during the passage of No Child Left Behind (NCLB), as policymakers sought to ensure all children access to "highly qualified" teachers. NCLB marked the most ambitious federal intervention into teacher preparation and regulation, yet in it legislators neatly ducked the substantial questions by largely deferring to state officials on how to define "highly qualified" teachers and how to identify them. It was not clear whether the "highly qualified teacher" provision was intended to encourage states to welcome teachers from multiple routes who met certain essential criteria, or whether it represented a victory by the "professionalizers" intended to endorse the role of teacher education and traditional professional groups. In the end, this studied ambiguity provided a comfortable refuge for legislators while raising the salience of the longstanding debate.

Today, amidst continued frustration that NCLB's overengineered apparatus, disenchantment with the legacy of the highly qualified teacher provision, and in an environment marked by visible enthusiasm for Teach for America and skepticism about the conventional efforts of teacher education and licensure, calls for radical rethinking have gained prominence. Especially evident among these are calls for using value-added assessment, promoting differentiated pay, and overhauling tenure.

Frustrated by efforts to change the culture of teaching through NCLB or the highly qualified teacher provision, would-be reformers turned to firmer measures. One popular tack has been the promotion of charter schooling. Another has been an attempt to circumvent state education officials, school districts, and teacher unions by finding new ways to push for dramatic changes in teacher evaluation and pay. The most notable such effort in the past decade was embedded in the Obama administration's Race to the Top program, which rewarded states for removing data firewalls, adopting performance-based pay, and revamping teacher tenure.

Now, value-added is an imperfect, imprecise, and fundamentally limited way to gauge teacher efficacy, but it's also a lot better than the status quo. The trick is to be honest and up-front about its limits and use it as tool that should be handled with care. Unfortunately, too often—as in Florida's Senate Bill 6 or the *L.A. Times* series—value-added is not used thoughtfully or carefully. Instead, it is frequently treated as a casual cure-all, leaving me nervous at how casually strong numbers on

Grades 3 to 8 reading and math value-added are treated as de facto determinations of "good" teaching. This tendency is amped up and energized by the frustration that reformers have had in getting professionalizers, education schools, or states and school districts to willingly embrace these tools in smarter and more deliberate ways.

The impatient rush to "fix" teacher quality in one furious burst of legislating is leading to troublesome overreaching and putting the cart before the horse. The result: hugely promising efforts to uproot outdated and stifling arrangements get enveloped in crudely drawn, sketchily considered, and potentially self-destructive efforts to mandate a heavy reliance upon value-added assessment.

Would-be reformers have trouble with the notion that unwinding a century's worth of accumulated policies needs to be a staged process. The first task is to uproot anachronistic policies and structures to create room for smart new solutions to take root. Only after a couple years in which we have given districts a chance to feel their way, and after a handful of alpha locales have crafted some promising approaches, does it make sense for state legislatures to start offering more direction. However, the resistance and folded arms of the professionalizers have led would-be reformers to overshoot and overprescribe in the fear that this is the only way they will change the status quo. Thus does half-baked, overly ambitious policy become the goal—simply because of the signal it sends.

Florida's Senate Bill 6, and some of what has been promoted under the banner of value-added, is a case in point. First, and least important, there's the question of small *n* sizes for many teachers and the challenge of devising good value-added assessment metrics for a raft of subjects. These challenges aren't necessarily susceptible to technical fixes and will require a little time, adjustments, and workaday compromises to take shape.

Second, systems built around individual value-added calculations can stifle the kind of smart personnel use that reformers are trying to encourage. Principals who have their best 3rd Grade reading teacher step in for grade-level colleagues, rotate faculty by strength, or augment classroom teachers with niche providers or online instruction are going to clash with a system predicated on evaluating how teachers are doing in their classroom with their students. If teachers are only teaching their kids 75% of the time, or if key instruction is being provided by others, then the results will be muddy.

Third, none of this matters yet. Florida's Senate Bill 6 went down over resistance to value-added calculations based on tests that didn't exist and that would not exist until the middle of the 2010s. Rather than using statute to impose systems that are currently nothing more than concepts, it makes more sense to strip out troubling mandates, create room for new solutions, and then build out the tests and data systems. This way, by the time those tools start to come online, we will have experience that can inform the deliberations of state lawmakers.

Right now, the best move would be for policymakers to push districts to base a substantial part of teacher evaluation and pay on measures of performance, for states to flag indicators deemed legitimate, and for districts, educators, and other providers to enjoy leeway in crafting and employing the specific metrics.

I understand the frustration with the status quo and union resistance that has fueled "fix it now" thinking. I understand fears that nothing much will change if states don't mandate it. But K–12 schooling is a big, complex exercise. Large, hurried solutions have a way of working less well than hoped. Impatience and lashing out in frustration can lead to bad policy—as with NCLB's 100% target for 2014 and its Kafkaesque remedy cascade. Both helped undercut support for the law's more sensible provisions and are now consigned to the trash heap. The patient path of welfare reform—create opportunities, nurture and monitor successful efforts, and then talk about new policy requirements—is the more promising one.

As I argued in the *Journal of Teacher Education* in 2005,

> The teacher preparation community does itself no favors by pretending that sharp critiques are necessarily malicious or mean-spirited. Arguing that preparation programs may be undemanding, ideologically biased, or less than rigorous in screening candidates is not out of line. If we are to have an honest and constructive debate, those skeptical of licensure or preparation programs must be able to question both institutional arrangements and the culture of teacher preparation without being vilified or excommunicated from the education fraternity. When even reasonable critiques are attacked as inappropriate, critics lose the incentive to self-police and it becomes difficult for the media, policymakers, or participants to distinguish hysterical critiques from serious ones.
>
> (Hess, 2005, p. 197)

The same point is true, perhaps more so, when it comes to teacher evaluation and pay. If professional educators, union leaders, and education school officials critical of and resistant to such notions would stipulate the good intentions of advocates and the desirability of rethinking familiar practice, they would find much firmer standing when it came to pointing out the excesses of Senate Bill 6 or the *L.A. Times* series, and would find a surprising number of new allies in their efforts. And, they might find more would-be reformers willing to treat them as full partners.

It is not merely that swapping epithets is aesthetically unpleasing. The larger point is that the vitriol tends to both stifle problem-solving and to be self-perpetuating. It shuts the door on fruitful debate and influences the rising genera-tion of advocates, thinkers, and practitioners. We create guarded camps that jeer at one another across the divide. In the end, this is neither democratic policy discourse nor even a thinking community; it is tribal politics. And it's not good for any of us.

So, as the teacher policy debates set forth on a new decade and we leave behind the "highly qualified teacher" and the debates of the NCLB era, it's time to make a choice. Will the next decade be another decade of line-in-the-sand declarations and dueling bouts of moral posturing, or will it be about how we use the mundane tools of data, termination, evaluation, markets, professional norms, training, and the rest to finally make teaching start to look like a profession in the modern era.

Reference

Hess, F. (2005, May/June). The predictable, but unpredictably personal, politics of teacher licensure. *Journal of Teacher Education 56*(3), 192–198. doi: 10.1177/0022487105275914.

FEDERAL POLITICS AND POLICY

A Misalignment of Incentives

Neal McCluskey

The U.S. Constitution is silent about education, and silence does not imply consent for federal action. Unless the Constitution explicitly says Washington may do something, the federal government may not do it. Federal forays into education illustrate why the Framers constructed the national government this way: Political incentives and policy logic rarely align.

Before illustrating this reality, it is necessary to establish that the Constitution does not, in fact, give the federal government authority to involve itself in education.

Article I, section 8 of the Constitution enumerates the specific powers given to the federal government, and among them you will not find "education," "school," or any related verbiage. Combined with the 10th Amendment—which reserves "powers not delegated to the United States by the Constitution" to the states or people—the federal government may not exercise authority in education.

But what of the beginning of Article I, section 8, which says that Congress has the power to "lay and collect taxes . . . to . . . provide for the . . . General Welfare of the United States"? Does that not authorize Washington to exercise power in education as long as it advances the general good?

No. James Madison—considered the central architect of the Constitution— refuted this interpretation in *Federalist* no. 41 (1787, see Rossiter, 1961), stating:

> For what purpose could the enumeration of particular powers be inserted, if these and all others were meant to be included in the preceding general power? Nothing is more natural nor common than first to use a general phrase, and then to explain and qualify it by a recital of particulars.

Providing for the general welfare only explains why the federal government is given its specific powers; the clause itself confers no authority. Similarly, the federal government is only authorized to "lay and collect taxes" to execute its limited powers.

So the federal government generally has no authority to make education policy. There are two limited exceptions to that.

First, the Constitution gives the federal government full jurisdiction over the District of Columbia, as well as all federal "Forts, Magazines, Arsenals, dock-Yards, and other needful Buildings." In those places, the federal government may involve itself in education.

Second, under the 14th Amendment Washington is given the duty to ensure that states do not deny to any person "the equal protection of the laws." As a result, the federal government must prohibit state discrimination in the provision of education.

Aside from those two things, Washington may not involve itself in education. Of course, from an education policy standpoint the crucial question is *why* the Framers so strictly limited the areas of federal concern.

The primary answer is that the Framers saw a national government as a considerable—but necessary—evil. A unified government was required to handle such inherently national duties as conducting foreign affairs and maintaining free trade between states, but it was also very dangerous. At the heart of the Framers' fears was the conviction that human beings are inherently self-interested, and if able will use government for their own, selfish ends. This threat comes from both political leaders and organized "factions" within society, and it is why the Constitution not only restricts the federal government to specific powers, but erects a system of checks and balances intended "to counteract ambition" with ambition (*Federalist* no. 51, see Rossiter, 1961).

Though Washington dabbled in education during the 19th and early 20th Centuries, it was not until the 1930s that arguably the most important check on federal overreaching fell apart. Thanks at least in part to President Franklin Roosevelt's "court packing" threat, which he launched after the U.S. Supreme Court had invalidated several New Deal programs, the Court stopped binding Washington to Article I, section 8. In *Helvering* v. *Davis* (1937), the Court rendered the enumeration of specific powers moot, ruling that the federal government could enact any law as long as it served the general welfare. That opened the door for federal officials and special interests—which Madison might have called factions—to use federal power for their own benefit, often without regard to educational logic.

This can be seen with the National Defense Education Act (NDEA). The NDEA, passed in 1958, was enacted in response to the successful Soviet launch of the *Sputnik* satellite. The rationale for the law was that the nation needed to greatly improve mathematics and science instruction so it could compete technologically with the Soviet Union. To a scared populace it might have seemed reasonable, but from a practical standpoint it made little sense. Even if teachers and students could be brought to much higher levels of performance by the law's limited teacher-training programs, the technological fruits would not be harvested for years. Had the Soviet Union truly posed an immediate technological threat to the United States, that time frame would have been much too long.

So why was the NDEA passed? Political calculations. It is now well established that President Dwight Eisenhower knew the nation was in little danger of falling behind the Soviets technologically (Payne, 1994). Nonetheless, he pushed many responses to the threat he thought unnecessary. He did so both to calm popular fears, and likely to prevent his party from facing the wrath of a frightened populace in the upcoming mid-term elections.

Eventually, *Sputnik*-induced hysteria disappeared and, with it, much of the support for federal intervention in education. Indeed, just a few years after NDEA was sped into law, a proposal by President John F. Kennedy to furnish federal funds for school construction and augmenting teacher salaries failed in Congress.

The situation changed again, though, in the mid-1960s, when President Lyndon Johnson's "Great Society" push yielded both the Elementary and Secondary Education Act (ESEA) and the Higher Education Act (HEA). Both laws, at their core, were focused on providing federal money to help low-income Americans access quality education.

From the standpoint of policymakers, it is hard to conclude that these laws were driven primarily by self-interest. Although there certainly could be political benefits from appearing to care about the poor—a positive public image, for instance—there is no clear, immediate benefit beyond that. The poor are not a sufficiently powerful voting bloc to seriously swing elections, and at the time the benefits of currying teacher and school administrator favor would have been limited. Most importantly, the National Education Association (NEA) did not become a union willing and able to mobilize large numbers of teachers for political action until about a decade later.

That said, educators did have an interest in getting the federal government involved in broadly funding education, and ESEA-friendly policymakers no doubt gained or maintained goodwill with them. Indeed, the NEA had lobbied for a greater federal education presence for decades, and in order to get the ESEA passed even dropped longstanding opposition to giving private schools access to some federal funds (Cross, 2004).

So, policymakers probably only had limited self-interest in seeing the ESEA passed, whereas educator groups had somewhat more interest. Even if both were being totally selfish, though, there was no evidence at the time that their interests weren't aligned with educational logic. Indeed, it is at least intuitively reasonable to think that providing greater resources to poor children would improve their educational outcomes. As the ESEA grew older, however, it became increasingly clear that that was not the case, with research generally revealing no positive effects. Yet the law not only lived on, it grew in financial size and scope. Why?

First, federal legislators could boast about spending money on education, and as ESEA evolved they broadened funding eligibility to touch more and more constituents. Policymakers were also providing money to educators and administrators, the people most motivated to be involved in education politics because their livelihoods depended on it. Unfortunately, the alignment of policymaker and educator interests rendered the law's poor educational effects largely irrelevant to policy decisions. As RAND Corporation researcher Milbrey McLaughlin concluded in a 1975 assessment of the ESEA, "the teachers, administrators, and others whose salaries are paid by Title I, or whose budgets are balanced by its funds, are, in practice, a more powerful constituency than those poor parents who are disillusioned by its unfulfilled promise" (1975, p. 71).

It was not until the 1980s that the self-interest of federal policymakers and educators started to diverge. The 1980 election, which brought conservative Ronald Reagan to the White House and swept a Republican majority into the Senate, was part of a national revolt against high taxes and government spending. Coupled with the *Sputnik*-like publication of *A Nation at Risk* in 1983, "standards and accountability" ultimately became the name of the game in education, leading to the erection of standards and testing regimes to which many educators objected. So, while politicians still had a keen interest in staying on the good side of education interests, they also had strong incentives to at least appear to hold schools accountable for poor performance.

For many years, major standards and accountability reforms occurred at the state level. But first with the Improving America's Schools Act—the 1994 reauthorization of the ESEA—and then with the No Child Left Behind Act (NCLB), federal education policy began to focus on "accountability" rather than compensatory funding for low-income students. Unfortunately, NCLB brilliantly illustrates what happens when political self-interest does not align with policy logic.

As policy, NCLB makes little sense. It is essentially a law divided against itself, making strict, sanctions-backed demands about standards, testing, and student proficiency, but leaving it to states to provide the substance of all those things. In so doing, NCLB all but forces states to set low standards but call them high, lest they and their districts be punished for kids not clearing proficiency bars.

From the standpoint of political self-interest, in contrast, NCLB is completely logical. The law enables federal politicians to appear tough on bad schools while simultaneously seeming dedicated to state and local control of education. It lets them be all things to all people.

The law's "highly qualified teacher" provisions are similarly self-interest driven and illogical as policy. By requiring that all students be taught by highly qualified teachers, NCLB sends the message to a concerned public that federal policymakers will no longer tolerate students being taught by, presumably, poor teachers. However, the law keeps state certification at the heart of teacher qualification while primarily just adding some testing hoops. It also makes no connection to student outcomes. That's not meaningful change if the assumption is that the status quo was hopelessly flawed, but it strikes an optimal political balance that helps policymakers look tough without overly angering powerful teachers' unions.

Of course, that politicians will act in their own interest is not restricted to the federal level. As the Framers well knew, self-interest is a human constant, and can manifest itself at any level of government. Still, the federal level is the most dangerous at which to vest power, both because it is the most distant from the people, and because if federal policymakers accumulate power and use it for their own ends one must leave the country for relief. That is why the Framers created a national government with few, specific powers, and encumbered by numerous checks and balances. In education, we have abandoned it to our detriment.

References

Cross, C. (2004). *Political education: National policy comes of age.* New York: Teachers College Press.

Helvering v. Davis, 301 U.S. 619 (1937).

McLaughlin, M. (1975). *Evaluation and reform: The Elementary and Secondary Education Act of 1965.* Cambridge, MA: Ballinger.

Payne, R. (1994). Public opinion and foreign policy threats: Eisenhower's response to Sputnik. *Armed Forces & Society 21*(1), 89–111.

Rossiter, C. (Ed.) (1961). *The federalist papers.* New York: Mentor.

EDITORS' COMMENTARY

The two essays in this chapter are written by individuals who offer different perspectives on teacher education policy. Frederick Hess, Director of Education Policy Studies at The American Enterprise Institute is a prolific writer on issues of education reform. Neal McCluskey is an associate director of The Cato Institute's Center for Educational Freedom with prior experience as a teacher and freelance reporter convering municipal government. Hess expresses dissatisfaction with the status quo while McCulskey turns to the U.S. Constitution to argue against a federal role in education. Except for protections guaranteed by the 14th Amendment, he asserts that vesting too much power at the federal level is dangerous "because it is the most distant from the people." This is a view shared by an increasingly vocal portion of the American electorate.

Hess offers a reflective description of the establishment of standards and licenses for teachers during the first half of the 20th Century. He refers to those who supported state teaching licenses—Education school deans, teacher education accreditors, teacher union representatives, and some state officials—as "professionalizers." The motives and actions of the professionalizers were unquestioned for a number of years until reformers—or critics of the status quo—began to find fault with public education. Hess argues that, instead of listening to the critics and trying to understand their concerns, the professionalizers countered that the answer was "more"—more funding to do more of the same. In the area of teacher education, that translated into additional pedagogical coursework. That response, Hess contends, made the reformers firmer in their resolve to go to great lengths to unmask the flaws in the education system. As he writes, "the resistance and folded arms of the professionalizers have led would-be reformers to overshoot and overprescribe in the fear that this is the only way they will change the status quo." The ultimate result was that many decision makers rejected suggestions by educators and dismissed research and studies done by them. We find evidence of this in the chapter recounting events in Florida.

Hess expresses frustration that years of federal interventions have not fulfilled policymarkers' expectations. Although McCluskey acknowledges that policymakers may have believed that targeting funds to schools serving children from low-income families through the Elementary and Secondary Education Act was a wise move, he notes that children living in poverty today do not seem to be much better off. In addition, McCluskey has reservations about powerful state teacher licensing laws with the apparent logic being that big state government is no more desirable than a big federal government.

Discussion Questions

1. Have the federal programs mentioned by Hess and McCluskey had an impact on school improvement or not? What are examples of federal programs that you believe have improved the education outcomes for children and youth?
2. Neal McCluskey acknowledges that the Constitution's 14th Amendment, which guarantees equal protection under the law, may be used to justify federal involvement in education. What are examples of education laws anchored to the argument of equal protection?
3. There are a number of individuals—including some educators and policymakers—who agree with these authors. Is there a middle ground where teacher educators and reformers could meet?
4. Frederick Hess finds fault with reformers for rushing to use value-added assessment measures before sufficient evidence is gathered on if they work, how they work, and when they work. How can the paradox of wanting an immediate solution and long-term research to see if an intervention will actually have the desired result be resolved?

3

DIMENSIONS OF TEACHER EDUCATION ACCOUNTABILITY

A Louisiana Perspective on Value-Added

M. Jayne Fleener and Patricia D. Exner

There has long been a movement in the U.S. for increased accountability of schools and programs that prepare teachers and school leaders. From a business model perspective, the simple idea that we should be able to measure the impact of a teacher on her students or the effectiveness of a teacher preparation program from the perspective of how well students of graduates from those programs perform has driven the development of state-wide data systems for measuring student achievement and connecting student growth over a year to their teachers and the teacher preparation programs that prepare them.

Many states, especially in the southeastern part of the United States, have been ahead of the pack of other states in developing state-wide student data systems that are necessary for predicting and comparing student growth over a period of time. The capability of tracking students and student performance and then linking that performance with teachers, comparing a variety of factors such as teacher experience, student or school poverty levels, and teacher preparation, requires complex data systems.

The U.S. Department of Education accelerated the development of state-wide assessment systems by providing financial incentives for states through grants and other competitive processes such as Race to the Top. In particular, the American Recovery and Reinvestment Act (ARRA), which includes the Race to the Top (RTTT) competition, emphasized the need for states to (1) establish data systems and use data to improve performance, and (2) increase teacher effectiveness and the equitable distribution of effective teachers. Systems focused on accomplishing these two objectives have been developed to produce evidence of effectiveness or the *value-added* of teachers and teacher preparation programs.

Louisiana has worked toward a value-added model for over ten years. Driven by teacher education reform efforts, the value-added approach is embedded in, and part of, a larger and more comprehensive assessment and accountability system for teacher education programs in the state.

We will discuss the development of the value-added model in Louisiana and the role colleges of education, especially those in research universities, may play in the future of teacher preparation and P-12 education. This is a story of changing expectations and shifting sands—the establishment of accountability expectations, the role of the state and university systems, and the challenge of *doing* teacher education.

The Louisiana Story

Over the past 11 years, there have been four major areas of emphasis on improving the effectiveness of teacher preparation in Louisiana. The agenda for reform of public and private teacher preparation programs in Louisiana began in 1999 with a state Blue Ribbon Commission for Teacher Quality, which mandated redesign of all teacher education programs based on more rigorous teacher certification requirements. As changes in certification requirements necessitated redesign of teacher preparation programs, in many cases new and more rigorous certification examinations were adopted with higher cut-off scores. The early stages of assessment and accountability measures for teacher preparation programs focused on outcome measures such as the number of program completers, numbers of university students pursuing certification in critical shortage areas, teacher candidate passage rates on national certification examinations, and numbers of graduates employed in state public schools in specified geographic areas with critical needs. Quality outputs were determined by teacher induction surveys completed by school principals and mentor teachers and by surveys completed by program completers themselves concerning their perceptions of the effectiveness of their preparation programs.

Because the redesign of teacher preparation programs, including significant changes in state licensure requirements, was the starting point for reform, an important part of the implementation of value-added measures in Louisiana included the continuum of policies and practices leading up to a comprehensive tracking and assessment system. Thus, the multiple layers and dimensions of teacher education reform in Louisiana have included the following four overlapping components (Burns, 2010):

- mandated redesign of all teacher preparation programs,
- required national accreditation,
- design of a new accountability system with performance scores calculated and published for each university preparing candidates for certification, and
- development of a value-added teacher preparation program assessment.

These four areas of reform intersect and interface at the level of colleges of education (see Figure 3.1). Like a pinwheel, colleges of education have become

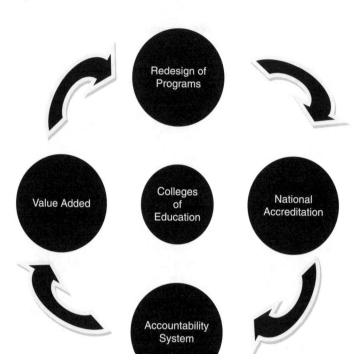

FIGURE 3.1

the sites or interstices of reform agendas that have placed new demands and challenges on, as well as opportunities for, these units. We will talk about each of these major shifts in the terrain of teacher education in Louisiana in order to present an analysis of the changing expectations of colleges of education in the U.S. We will present the position that each of these perspectives embeds and implicates colleges of education as ambiguously located between P-12 and higher education. The at times uncomfortable positioning of colleges of education in the in-between spaces presents both an opening for reinvention and a closing of traditional core missions. Value-added assessment, as part of the teacher preparation accountability system, presents an entirely new dimension for universities preparing teachers.

Redesigning Teacher Preparation

Beginning in 1999, driven by the desire and need to address increased public demands for accountability of teacher education programs, the Louisiana state policymaking agencies that oversee public higher education, P-12 schools, and teacher preparation programs formed the Blue Ribbon Commission for Teacher Quality. In Louisiana, teacher preparation programs in public institutions have many masters, including the

Board of Regents for Higher Education, the Board of Elementary and Secondary Education, and the Louisiana State Department of Education. The formation of the Blue Ribbon Commission was important for coordinating and setting policy affecting the preparation of teachers and the accreditation of teacher preparation units in the state.

The Blue Ribbon Commission was constituted to include 36 board members representing state, university, district, school, teacher, community, and parent constituencies. State public and private universities each had a representative. Arts and Sciences faculty were also represented on the Commission, as were principals, teachers of the year, board certified teachers, district superintendents and human resource representatives, and the state Parent-Teacher Association (PTA). The Blue Ribbon Commission was co-chaired by a member of the Board for Elementary and Secondary Education (BESE) and a member of the Board of Regents, with the goal of ensuring that teacher certification and preparation, under the purview of BESE and the Louisiana Department of Education, and university programs and degrees, under the province of colleges and universities within the Regents of Higher Education system, would coordinate as reforms were negotiated, approved, and implemented.

With the passage of the No Child Left Behind Act in 2001 (U.S. Department of Education, 2001), the agenda for the Blue Ribbon Commission addressed specific aspects of Title II, Part A, *Preparing, Training and Recruiting High-Quality Teachers*, emphasizing "the need to prepare, train, and recruit highly qualified teachers and principals in order to have a positive impact on student achievement" (Louisiana Department of Education, 2010). According to Dr. Jeanne Burns, the Special Assistant to the Governor and Board of Regents staff facilitator for the Blue Ribbon Commission, the Commission served as "the button that allowed [Louisiana] to develop a system" (Exner, 2001) of teacher preparation program accountability. This button was not only the button launching the rocket ship of value-added assessments for the state, but was also the panic button for many teacher preparation programs in Louisiana.

Early on, the deans of education found the Blue Ribbon Commission agenda to be objectionable. In particular, while input for mandated program revisions was solicited from the deans, the deans shared the general consensus that their perspectives were not accommodated in subsequent program redesign components and approval processes. A feeling of disenfranchisement was exacerbated early on by the lack of voice of the deans on the Blue Ribbon Commission and the limited opportunities for public comment during the Blue Ribbon deliberations. The leadership challenge for deans was minimizing and managing tensions created by presenting redesign mandates to faculty and university administrators, while at the same time encouraging faculty to participate and engage in the redesign process and overseeing their efforts. In these early stages of the redesign process, college of education deans often found themselves in conflict with their faculty, as well as with those to whom they reported for accreditation and program approvals, including the Louisiana State Department of Education, the Board of Regents, the Board for Elementary and Secondary Education and their own campus provosts. The deans were often stigmatized as "unwilling to change" and "obstacles to reform" by the policymakers pushing the reforms and as "dictators" and "poor representatives of academic freedom" by beleaguered faculty growing weary of redesign mandates. Consequently, many of the college of education deans left their positions during the years between 2001 and 2005.

The pressure for colleges of education during the redesign of programs was further complicated by the multiple years of program revision that occurred, from the university perspective, in a piecemeal fashion. Rather than providing scaffolding for changes across all program areas to be developed and implemented over a short period of time, the Commission redesigned different program areas and created new ones throughout a period of seven years, with the first years marked by often monthly changes in redesign guidelines. So, for example, in the first year, the programs which underwent redesign and the rigors of an approval process that included bringing in external consultants were elementary and secondary education, plus the addition of a new certification program in Grades PK-3. After seven years of redesigning programs, which included P-12 and several advanced master's programs, faculty were worn out and students were confused as old programs were phased out and new programs were implemented. Faculty were frustrated further by efforts of colleges of education to go through existing campus processes for new program approvals when oftentimes, programs were approved by the various governing bodies before campus committees had been engaged in the process. In many cases, programs were approved by the Board of Regents as new programs offered by the universities before the university processes had been played out.

Accreditation and Accountability

The second and third aspects of developing a comprehensive assessment system for teacher preparation programs in Louisiana's were re-examining accreditation requirements and developing a system of accountability that involved multiple measures including value-added. Louisiana's changes to teacher preparation program expectations thus included the expectation of national accreditation of all Louisiana university-based teacher preparation programs by the National Council for Accreditation of Teacher Education (NCATE). (Later the Teacher Education Accreditation Council (TEAC) was added as an accreditation option.) The intent of and goal for requiring national accreditation was to ensure that redesigned programs met and continued to address national as well as state expectations for teacher preparation. At the same time national accrediting agencies were also changing their expectations for demonstrating the quality of programs by focusing on performance-based systems of accreditation. While some Louisiana universities had long participated in national accreditation processes, the expectations for accreditation of all universities had thus shifted for all institutions to national accrediting agencies at the same time that national expectations were shifting from conceptually based to performance-based evidence.

The accreditation mandates were generally more acceptable to the teacher education community in Louisiana than were the requirements for redesign of programs. There was among some faculty a sense that national accreditation provided evidence of quality that was not as clearly apparent in mandated redesign programs, especially to faculty who felt their traditional programs were superior to the newly created programs. There was, however, through the accreditation process, increased understanding of the relationship among program coherence, planned and systemic performance assessments, demonstration of the quality of programs, and accreditation. Thus, increased emphasis on performance-based evidence for national accreditation reinforced redesign components as intimately connected with assessment processes.

Nevertheless, the burden of accreditation on programs that were in flux with redesigned components in various stages of approval and implementation intensified the anxieties of going through the accreditation process with new standards and higher expectations for demonstrating quality through candidate data. The gulf between teacher education faculty and non-teacher education faculty in colleges of education widened at many institutions, with differing and increased demands on teacher education faculty to participate in the redesign and accreditation processes. Especially in research extensive institutions, teacher education programs were increasingly staffed by non-tenure track faculty or early career, non-tenured faculty. Teacher education research has in many cases shifted outside of the academy and, lacking research-active faculty among the teacher education faculty ranks, the "second-class" status of teacher education has further diminished the value universities place on teacher preparation.

Accreditation has driven assessment and accountability in Louisiana, not only with the emphasis on performance standards, but with the creation, through the Blue Ribbon Commission, of a comprehensive, outcomes-based accountability system that includes value-added measures. Underlying questions driving the assessment and accountability system in Louisiana include the following:

- How do we link student performance to teachers and the programs that prepare them?
- How do we control for the many variables that contribute to, and distract from, a student's ability to learn?
- How do we control for the many variables that contribute to the effectiveness of beginning teachers?

These are just some of the myriad challenges presented by the assessment and accountability movement that supported the creation and adoption of the value-added approach to assessing teacher education programs in Louisiana.

The Value-Added Dimension

Louisiana was positioned to become a leader in value-added assessment of teacher preparation programs because of the comprehensive and systemic approach taken by the Blue Ribbon Commission for reforming teacher preparation programs and establishing the accountability of these programs. Also contributing to the development of the value-added system of assessing teacher preparation programs was a student data system already in place that could uniquely identify each student in a teacher's class and track previous performance of the student in tested grades. At the school level, No Child Left Behind (NCLB) mandated that all public schools in the U.S. use standardized tests to measure student achievement. According to the Quality Counts 2010 Report (Quality Counts, 2010, pp. 33–49), all 50 states and the District of Columbia had in place multiple choice assessments to measure student performance. Missing in many states, however, was the capacity to link student scores across years and to link current and past scores with individual teachers. These two components were integral to the development of the value-added assessments in Louisiana.

The phrase "value-added" refers to the ability to quantify the contribution teachers, schools, districts, or teacher preparation programs make toward the achievement or growth of students. While NCLB provided for schools to demonstrate year-by-year changes in aggregate performance of students in a school or district, value-added effects demand more than comparing this year's 4th-graders with last year's 4th-graders. Value-added measures require tracking student performance across years in order to predict and compare expected and actual learning gains. Louisiana was among the first to extend the notion of value-added analyses to assess the contribution teacher preparation makes to student achievement. This perspective of value-added teacher preparation programs in Louisiana had its origins in the 2002 White House Conference on *Preparing Tomorrow's Teachers* (Whitehurst, 2002).

Russ Whitehurst, the director of the Institute of Education Sciences (IES) under the George W. Bush administration in 2002, questioned the relationship between student learning and teacher preparation, arguing the dearth of evidence for the positive effect of certification status of a teacher on student achievement (see Whitehurst, 2002). His criticism of teacher preparation programs as necessary for beginning teacher effectiveness opened the door for multiple pathways to teaching and refocused value-added analyses for evaluating teachers to evaluating teacher education programs. According to his review of the research literature, teacher certification status per se was not an important factor. Variables like subject matter knowledge, teacher verbal and cognitive ability, and classroom experience, according to his analyses, seem to be supported by the research as relevant and important factors in student achievement. Certification and preparation status— for example, whether a beginning teacher went through a teacher certification program or not—was not, according to Whitehurst, a factor for which there is compelling evidence of a positive effect.

In this context, Louisiana's approach to value-added assessments focused on teacher preparation programs rather than on individual teachers. Developed by George Noell of Louisiana State University Department of Psychology in the College of Arts and Sciences (see Noell, 2005), the Value Added Assessment of Teacher Preparation Programs (VAA-TPP) model was developed to become a component of the teacher preparation accountability system.

> The Value Added Teacher Preparation Assessment predicts growth of student achievement based on prior achievement, demographics, and attendance, assesses actual student achievement, and calculates effect estimates that identify the degree to which students taught by new teachers showed achievement similar to students taught by experienced teachers. The teacher preparation effect estimates are based upon multiple new teachers in multiple schools across multiple school districts in the state. The predictors examine student variables, teacher variables, and building variables and differ slightly based upon the five content areas examined which are mathematics, science, social studies, reading, and English/language arts.
>
> (Burns, 2010, p. 5)

The Louisiana VAA-TPP is different from other value-added models in at least four ways. First, rather than providing effect scores for individual teachers, the Louisiana VAA-TPP provides effect scores for programs that prepare new teachers. The VAA-TPP was not designed for, nor validated to provide, individual teacher effect scores.

Second, teacher preparation program effect scores are calculated for the aggregate of first- and second-year teachers prepared by a program and compared across programs and with experienced teachers. Thus, effect scores compare redesigned teacher preparation programs based on analysis of first- and second-year teachers from the programs and that of aggregates of experienced teachers and other new teachers.

A third difference of the Louisiana VAA-TPP from other value-added models is that, because scores are aggregates of the performance of students of new teachers from different preparation programs in different content areas, different effect scores are calculated for the five content areas tested in Louisiana (mathematics, science, social studies, reading, and English/language arts). So rather than receiving one effect score, teacher preparation programs receive up to five scores for each testing area in which they have sufficient program completer numbers.

A fourth characteristic of the Louisiana VAA-TPP is that teacher preparation programs are categorized into one of five "bands" depending on how their first- and second-year teachers compare with experienced teachers and/or other beginning teachers from other institutions across the state. The categorical per-formance bands assigned to teacher preparation programs in the areas of mathematics, science, social studies, English/language arts, and reading in Grades 4–9 public schools in Louisiana are as follows.

I.   Programs for which there is evidence (although not necessarily a statistically significant difference) that new teachers are more effective than experienced teachers.
II. Programs whose effect is more similar to experienced teachers than new teachers.
III. Programs whose effect is typical of new teachers.
IV. Programs for which there is evidence that new teachers are less effective than average new teachers, but the difference is not statistically significant.
V.   Programs whose effect estimate is statistically significantly below the mean for new teachers.

Effect scores for programs performing at levels I–IV are not statistically significant. These scores, reported by performance level bands for each of the five content areas in which a teacher preparation program has sufficient program completers, are made public. Thus, the VAA-TPP has become part of the public accountability system for teacher preparation programs and will be factored into the teacher preparation program assessment and accountability model.

The VAA-TPP model in Louisiana relies on and is made possible because of the Louisiana Educational Accountability Data System (LEADS), a comprehensive data system that links students to teachers and links to other testing and supplemental databases that provide information about student attendance, enrollment, disability status, free lunch status, race, and gender, as well as teacher preparation information. In order to be included in the analysis, students have to be continuously enrolled in a teacher's class from September 15 to March 15. Other factors (such as taking the same end-of-year assessment over two years rather than matriculating to the next year test) are considered in the analysis and decision whether to keep students in the class population used to calculate teacher and teacher preparation effect scores. Likewise, students with more than one teacher in a particular content area are included in the analyses for each teacher with a reduced weighting.

The Louisiana VAA-TPP includes in the calculation of unit effect scores only teacher preparation program graduates who meet these specific criteria. Thus, not all graduates from a teacher preparation program may be included in the calculation of the preparation program effect score. As described in the VAA-TPP Technical Report 2009 provided to teacher preparation units[1]:

> For purposes of illustration, assume that a TPP had 100 graduates in a particular year. Of these graduates, some will teach subjects such as band, foreign language, or physical education. Assuming that 20% of the graduates teach in these areas, 80 new teachers would remain whose effects on student

achievement theoretically could be estimated. Of the 80 new teachers, some will not enter public school teaching. They will teach in private schools, pursue graduate study, delay work entry to start families, or pursue employment outside schools. This part of the attrition could readily reduce the number of available new teachers to 50. Of this number, half will typically teach outside tested grades and half will teach in tested grades. Of this 25, assume approximately 13 teachers teach all subjects in the elementary grades and 12 teach a single content in middle school or high school (i.e., 3 teachers per content area). If this pattern held, there would be 16 teachers per content area in each year's cohort. The assessment model capitalizes observations of teachers across three years, so in this assessment, two graduate cohorts would be required for the TPP to be included in the analysis.

So in this scenario, if a teacher preparation program had 100 graduates in a year, only 16 might actually count in the calculation of the program effect score, requiring two years or more before the institution's graduates are included in the analysis and provided for feedback to the universities and for accountability purposes.

Preliminary findings of the value-added analyses of teacher preparation programs in Louisiana suggested that the performance of first- and second-year teachers varied not by institution but by content area. Thus, for example, one teacher preparation program may have graduates who, as an aggregate, performed better than graduates from other programs in mathematics, but who did not do as well in language arts or science. There are many questions raised by the use of value-added approaches to assess the effectiveness of teacher education programs, especially related to reporting procedures.

Challenges to Value-Added Models

In her critique of value-added models that assess teachers, Audrey Amrein-Beardsley cited three basic limitations (Amrein-Beardsley, 2009, pp. 38–42):

1.   Reliance on Standardized Tests
2.   Lack of Evidence of Validity
3.   Lack of Transparency

These same criticisms can be levied against the Louisiana value-added model that assesses effects of teacher preparation programs.

Although the Louisiana VAA-TPP relies on student achievement scores along with other variables that the model has determined to be factors, the issue of overreliance on a single measure of achievement is a problem for the Louisiana model. Linking students to teachers by comparing the history of performance on standardized tests while factoring in student attendance, enrollment, disability status, free lunch status, race, gender, and teacher preparation program still reduces student growth to achievement test performance. As Amrein-Beardsley argues, "It's unclear whether standardized tests can accurately measure what students know and are able to do at one point in time, let alone over time to measure 'knowledge added'" (2009, p. 39). This limitation is confounded by the fact that only certain grades are tested (Grades 4–9) and the tests from year to year may not be equivalent with regard to content coverage on the actual exam.

The reliance on standardized tests for the Louisiana VAA-TPP introduces a different set of obstacles while solving some of the more traditional concerns. Through a complicated data cleansing process, issues of missed tests, student mobility, teacher mobility, teaching assignment, and teacher certification area have been addressed in the Louisiana model. Furthermore, by aggregating effects of all beginning teachers from a particular teacher preparation program to calculate preparation effect scores, individual student or teacher abnormalities in testing can be statistically controlled. And finally, with regard to the issue of single measures of accountability, the value-added effect scores are just one of multiple metrics used to assess teacher preparation programs. What is lost in the Louisiana analyses is the ability to drill down to determine which of the approximately 16% of teachers from any given program are used in the calculations. Are these our best students? Where are they teaching? Were they elementary or secondary certified? Many of these questions lead to concerns of validity.

Traditionally, value-added studies are associated with informing administrators of teacher performance that can be used in annual evaluations, reward incentives, future classroom teaching assignments, or renewal of contracts. Likewise, value-added statistics have been used to predict future student success or challenges, suggesting strategies for addressing achievement gaps and preparing students for postsecondary experiences. The Louisiana VAA-TPP model is not valid, nor does it claim to be valid, for making these kinds of decisions or predictions. What it does claim to do, however, is to provide a basis for categorizing teacher preparation programs according to how well their beginning teachers perform. The challenges to validity of the Louisiana VAA-TPP model are in ranking programs in categories I–V when levels I–IV are not statistically significantly different. A college can be ranked as a level III, for example, suggesting their new teachers perform as well as other new teachers from other programs across the state, yet only be 1.5 points below the mean of 300 of all teachers in the state. It is not clear what an effect score of −1.5 means both in terms of actual effects and long-term effects for student success.

University administrators and deans of colleges of education in Louisiana are presented with results of programs prior to public release of the analyses. University system presidents and vice presidents are first briefed on the rankings of all preparation programs in the state, and then campus administrators, including chancellors, provosts, and deans of education, are presented with state-wide results before the analysis is made public. Deans are expected to respond to their results, first internally across campuses and with their faculty, then externally as results are made public. In one particular case, a university received national attention because one of their traditional undergraduate teacher preparation programs was ranked at level IV, suggesting that their beginning teachers, on average, do not perform as well as other beginning teachers across the state. Although the effect scores were not significantly different from programs with levels I–III designation, the dean and other upper level administrators responded proactively by seeking additional information from the VAA-TPP staff so that program analyses and adjustments could be made. Their efforts to use the VAA-TPP data for program improvement received national attention, and their response was used to support the use of these kinds of value-added analyses. The question remains, however, what an effect score of, say, −10 on a standardized score of 300 really means, especially given that the score is not statistically different from an effect score designating level I–III performance.

Unless and until mechanisms to provide useful information for program improvement are provided, the value-added designations have little usefulness. Does ranking of a teacher preparation program at a level III in mathematics with an effect score of −1.5, for example, point to needed changes in the elementary mathematics course sequence, or are the teachers that brought the scores down secondary teachers teaching at Grades 6–9, suggesting perhaps the need to change methods courses for secondary teachers to teach in middle school? Is the issue of scores in any given year skewed by the kinds of schools in which our beginning teachers are placed? Or are the results skewed by which students from a particular teacher preparation program were included in the analyses? In the case of the university receiving level IV designation, some additional data were provided to allow for further analyses. But until data at the individual teacher level are provided to teacher preparation programs, informed program adjustments and helpful analyses for interpreting the performance of beginning teachers are limited. What remains is a system that feels punitive to colleges of education. The expectations for public accountability are driving reporting strategies, placing pressures on teacher preparation programs to address potential perceptions that their programs are producing inferior beginning teachers despite the lack of statistical significance or even representativeness of a majority of graduates from a program being included in the analyses.

These validity questions lead to issues of transparency. Without additional data down to the individual teacher level, the model cannot be tested and the results cannot be interpreted for program adjustments. As the model changes according to factors that are found to be relevant to student scores, researchers and colleges of education need to have access to these data in order to make independent determinations of the appropriateness of selections of key variables. The researchers running the value-added model argue that the model is not valid at the individual teacher level, but providing university program researchers data at just such an individual level is necessary in order to interpret the meaning of the results for program adjustments to be made. Without knowing the specific teachers whose scores are used to calculate a preparation program's value-added effect, and because as few as 16% of the graduates may be included in any given year, there is no way to know if the teachers are representative of that cohort of graduates or what characteristics of these teachers may have affected results.

It could be that there are systemic inequities exacerbated by the care taken to exclude teachers who do not meet certain criteria. For example, what can be concluded about a teacher education program if graduates who meet all of the criteria to be included in the value-added analyses are primarily teaching in high achieving schools, while graduates in low achieving schools are disproportionately eliminated from the analyses because of the data cleansing strategies, especially if teacher preparation programs are not given the data to make these interpretations? Without transparency of the data, researchers and preparation programs that may have a better understanding of the data cannot verify the reliability of the population. Furthermore, in states like Louisiana, where teacher and student mobility may be more the norm than the exception, excluding these teachers in the calculations may introduce an unrealistic picture of effective teaching in the state.

There is a larger question about value-added measures of teacher preparation programs that emerges, however, beyond these questions of validity and appropriateness of using standardized tests. This question pertains to the role of teacher education, especially in research universities. In an era of value-added assessment and high stakes accountability, do we lose the ability to innovate, experiment with, and adapt teacher education to meet the changing needs of our students? Does teaching become relegated to the mechanics and skills of curriculum delivery, leaving little room for the art (and heart) of teaching that touches kids in ways that are not measured once a year on a standardized test? Do the demands for both teachers and teacher preparation programs to teach to the curriculum remove the unique role that colleges of education have to shape a profession, including challenging young minds to think outside of the box? Do issues of diversity and personal philosophies of education have any place in teacher preparation programs housed in universities if teaching effectiveness and program quality are predominantly measured by value-added statistics which do not capture the multiple dimensions of the art of teaching?

The Future of Teacher Education in Universities

The Louisiana case of teacher education reform over the past 11 years is a study in, and a challenge to, the role played by colleges of education in both the P-12 and university settings. For most of the 20th century, as stand-alone normal schools were eliminated, education colleges became part of comprehensive university offerings. The preparation of teachers and other P-12 professionals in comprehensive universities carried the demands of academic as well as vocational or professional preparation. Expectations of education students, as well as of their faculty, were to shape, inform, and advance the education system, not simply replicate it. Colleges of education in research universities, in particular, were called upon to research and inform practices to challenge the status quo and adapt P-12 education to meet the challenges of the future.

As expectations for professional development schools and school partnerships increased in the 1980s and 1990s, colleges of education became more and more a bridge between P-12 and higher education. As a bridge, colleges of education are expected to provide valuable resources and opportunities for both P-12 and university partners to support pathways to postsecondary education and job-readiness. Colleges of education have been called upon to provide in-service training and continued support to classroom teachers, while also serving as a conduit for communications and programs offered by university peers. Added to these expectations are increasing demands that education research focus on solution paths for some of the most intractable and complicated problems associated with P-12 education,[2] while Education schools, and their students, are evaluated on how well P-12 students perform on standardized tests.

The positioning of colleges of education as a bridge between P-12 and higher education introduces ambiguities to core mission and purpose. As Figure 3.2 depicts, colleges of education occupy what has become more and more the uncomfortable ground between P-12 and higher education—not fully occupying either space.

The complicated relationships colleges of education have with their P-12 and university communities have raised challenges to the very existence of, and need for, colleges of education. This uncomfortable existence is exacerbated by value-added studies that tie teacher preparation to P-12 student performance and by increasing public demands for alternative pathways to teaching. The conundrums faced by colleges of education suggest the need to revise goals and operational understandings. If we are to occupy the in-between spaces of P-12 and higher education, we need operational capabilities that allow us to refocus core missions and purposes. A fundamental property of any living or social system is to survive.

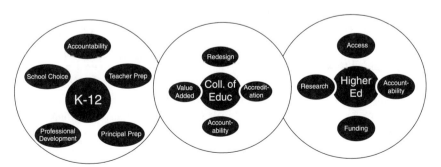

FIGURE 3.2

It is time colleges of education reinvent themselves with an understanding and appreciation for our role as a bridge between P-12 and higher education. Value-added accountability plays an important role in how we need to revision our relationships across the academic worlds. If we do not reimagine and recreate our relationships among P-12 and higher education, we will become irrelevant to the preparation of the next generation of teachers.

Emerging from biological metaphors, complexity theory is "a theory of change, development and evolution through relationships" (Morrison, 2008, p. 22). Viable complex systems creatively and synergistically accommodate and adapt to changes and challenges in their environment and as they grow and change throughout their lifespan. Key characteristics of complex systems capable of adapting to rapid change and growth include *clear purpose, operational functionality, relational openness, and environmental connectedness*. Review of these characteristics as related to systems such as colleges of education and the impact of value-added measures can be quite revealing as to our future.

The *purpose* of complex adaptive systems at both organismic and social systems levels is to respond to environmental challenges in ways that maintain system identity. Often described in terms of self-organization, complex adaptive systems have the ability to creatively respond to change while maintaining identity through centrality of purpose and mission. What is the purpose of colleges of education? Have the mission, purpose, and goals become so diffuse and so defined by others that the field of education itself has no meaning? What is the impact of value-added accountability on changing our core mission and goals? How the purpose of an organization is enacted is evident through its operational functionality.

Operational functionality in complex adaptive systems includes dispersed knowledge throughout the system, internal integrity as action remains coherent with system purpose, environmental openness to interact with and understand environmental change, and cooperation rather than competition so actions by the system maintain the health of the environment as well the health of the system. Operational functionality embraces values that include parameters for action in order to fulfill system purpose. Peter Senge studied businesses as complex adaptive or learning systems (Senge, 1990), offering numerous examples of businesses where shared vision and goals were enacted independently, at all levels of the organization, without top-down control or mandates, as agents within the system made decisions based on values consistent with and in order to maintain the shared vision. Providing first-class service to customers, for example, is a value that drives decisions and action at all levels of a business with the vision of expanding to all segments of a consumer population by meeting their technological needs. This dispersed knowledge of business goals and operational understandings allows individuals to act in ways that are consistent with the purpose of the organization but that take into consideration situational factors requiring unique responses. Senge especially found complex adaptive systems, or learning organizations, to be a useful way to understand successful organizations in a knowledge economy. In non-human animal contexts, a beehive is often used as an example of a complex adaptive system in which independent agents within the system behave autonomously but contribute to the growth and wellbeing of the collective whole. Healthy responses within a complex system do not require centralized control or uniformity of action but maintain system wellbeing precisely because independent agents within the system act autonomously and uniquely while being driven by common purpose and goals for individual and system sustainability. Yet again, have the mission, purpose, and goals of colleges of education become so tightly defined by demands for uniformity that such healthy responses are disabled? How do we exist in the in-between spaces of P-12 and higher education in an environment of increasing accountability if we do not have the autonomy or will to repurpose our efforts? Do we have a role to play in research and policy affecting P-12 and teacher education? What should our research missions be? How do value-added accountability measures impact our mission to prepare teachers both for meeting today's challenges as well as for shaping and addressing the unknown challenges of tomorrow? How do externally mandated policies affect shared understandings and core values of colleges of education? How do we intelligently respond to the sometimes hostile environment of accountability?

Relational openness in complex adaptive systems entails continual monitoring of environmental factors that may have an impact on the system. In complex social organizations, there are multiple layers and levels of relational openness both within the system and by agents within the system with the external environment. Relational openness is necessary for intelligent action as a system adapts to changes and challenges in environmental contexts. Can colleges of education adapt to such changes and challenges, given mandates from multiple levels? Are missions and goals so diffuse that intelligent action is no longer even possible?

For complex organizations, relational openness is often difficult either because changing operational functionality is too difficult or because mechanisms to receive feedback from the environment have not been established. Accommodating feedback from the environment is a function of *environmental connectedness*. Positive as well as negative feedback are important for complex adaptive systems to take information from the environment and respond appropriately. While negative feedback serves as a damper to resist willy-nilly change, positive feedback provides for emergent, creative change. Often referred to as the snowball or butterfly effect, positive feedback can accommodate small perturbations resulting in dramatic and fundamental changes to the system. Are colleges of education receiving both types of feedback, allowing for action rather than simply reaction? What kind of data from value-added studies are necessary in order for colleges of education to be able to be environmentally connected, in a position to use the data for adaptive responses?

How might treating colleges of education as complex adaptive systems provide opportunities for creative and dynamic responses to the challenges of education? When considered in the context of complexity theory, value-added studies of teacher education represent and exacerbate the conundrum of teacher education in higher education. As colleges of education continue to serve their many masters, complying with multiple and sometimes conflicting mandates for uniform programs, metrics, and goals, adaptive potential becomes limited. Like the chicken with its head cut off, these systems may survive for a while, but they will eventually die. Fundamental system intelligence has been eliminated when colleges of education lack key information upon which future strategic action can be based. Without this knowledge, and lacking autonomy and clarity of purpose, colleges of education will become nothing more than an appendage of either P-12 or higher education. Without the ability to interact and engage in creative and synergistic ways with P-12 and higher education, the heart and soul of colleges of education are being replaced by assembly-line mechanics. Colleges of education are becoming cogs in a machine rather than partners in a process. Focusing on achievement through value-added studies has, up to this point, curtailed researching and exploring other aspects of student growth and narrowed the focus of education to "adequate yearly progress" of students (and "schools") as measured by achievement tests.

In order to better serve the current and future needs of higher education and P-12 education, colleges of education need to clarify their purpose and, in so doing, determine an identity that sets them apart. Colleges of education have lost their identities, especially as they have become merged in university units with a variety of areas including kinesiology, social work, and nursing. Our contribution to the knowledge domain of education is especially lacking an identity. As a highly interdisciplinary field, we borrow from psychology, sociology, anthropology, neurology, and ecology (see Davis & Sumara, 2006). While interdisciplinarity is a strength of educational research and disciplinary action, we need to be clear about the differences between what we do and what these other fields of study do. We need to clarify the kinds of knowledge and action that feed back into the system of education that becomes our unique contribution to the adaptation and sustainability of P-20 systems of education. We need to be clear about the operational functionalities, including the sources and limits of necessary resources and funding, in order to strive toward our purpose(s) and goals. If we are driven by the values of meeting the educative needs of children, for example, how might we build into our system of colleges of education adaptive processes whereby independent and creative action is nurtured rather than squelched; where coherence of theory and action, research and teaching, individual and collective learning is supported; and where collaborative partnerships are synergistic?

Value-added studies and the push for information about teacher effectiveness are not the bad guys in redefining colleges of education, but are publicly symptomatic of our own lack of clarity of purpose, role, values, and relationship to P-12 and higher education. Colleges of education, like the chicken, will eventually die if we do not re-establish the perspective that distinguishes who we are in the changing P-20 educational landscape.

The Louisiana perspective on value-added accountability and the history of the development of systems of accountability for teacher preparation in Louisiana are symptomatic of our ambiguous role in between P-12 and higher education. We have opportunities to reinvent who we are and what we stand for, not in opposition to our standing as a bridge, but in appreciation for the unique terrain that we have occupied since our inception. Value-added research has not changed our position in the university, but has brought to light that our natural and adaptive advantages have not been served by trying to be like engineering or the humanities. We are uniquely positioned to shape the future by touching the lives of children. This is an awesome challenge and responsibility.

Notes

1 No public archive of this report can be found on either the BESE or Board of Regents websites.
2 See, for example, Easton (2010).

References

Amrein-Beardsley, A. (2009). Value-added tests: Buyer, be aware. *Educational Leadership, 67*(3), 38–42.

Burns, J. (2010). Testimony for U.S. House of Representatives. April 20. http://edlabor. house.gov/hearings/2010/05/supporting-americas-educators.shtml

Davis, B. & Sumara, D. (2006). *Complexity and education: Inquiries into learning, teaching, and research*. Mahwah, NJ: Lawrence Erlbaum Associates.

Easton, J.Q. (2010). Out of the tower, into the schools: How new IES goals will reshape researcher roles. Presidential Address, American Educational Research Association Conference, Denver, CO, May 2. Retrieved May 30, 2010 from http://ies.ed.gov/ director/biography.asp

Exner, P.D. (2001). Teacher education reform and accountability: A study of the Louisiana Blue Ribbon Commission on teacher quality. Dissertation, Louisiana State University.

Louisiana Department of Education. (2010). Retrieved May 30, 2010 from http://www. louisianaschools.net/lde/pd/1045.html

Morrison, K. (2008). Educational philosophy and the challenge of complexity theory. *Educational Philosophy and Theory, 40*, pp. 19–34. doi: 10.1111/j.1469–5812.2007.00394.x

Noell, G.H. (2005). *Assessing teacher preparation program effectiveness: A pilot examination of value-added approaches II*. Originally retrieved from http://asa.regents.state.la.us/TE/ value_added_model

Quality Counts (2010). State of the States: Holding all states to high standards, *Education Week, 29* (January 17).

Senge, P. (1990). *The fifth discipline: The art and practice of the learning organization*, New York: Doubleday.

U.S. Department of Education. (2001). *No Child Left Behind Act of 2001,* passed by the U.S. House of Representatives on May 23, 2001, the U.S. Senate on June 14, 2001, and signed into law by President Bush on January 8, 2002. www.ed.gov

Whitehurst, G.J. (2002). Scientifically based research on teacher quality: Research on teacher preparation and professional development. White House Conference on Preparing Tomorrow's Teachers, March 5, 2002. http://www2.ed.gov/admins/ tchrqual/learn/preparingteachersconference/whitehurst.html

EDITORS' COMMENTARY

This case is about the responsibility of teacher educators to the state's K-12 schools. Is the teacher education program an integral part of the state's responsibilities to serve the public schools of the state? If so, does the teacher education program have responsibilities to the state and its schools that trump the claims of traditional academic autonomy and independence by the collegiate program? It can be argued that education school faculty have a tripartite responsibility to the state, the institution, and the profession. As such, part of the responsibility of being a teacher educator is to conform to the demands of the state relative to its expectations and the needs for an adequate supply of beginning teachers capable of teaching the state's K-12 curriculum. Consequently, traditional faculty prerogatives of academic freedom are overridden by the compact between the state and the approved teacher education program. This case suggests that state expectations often trump academic autonomy when it comes to preparing teachers.

A second theme that emerges in this case is the uncertainty of the role that teacher education should play between the university and the school. Does teacher education and/or the education school occupy that space and, therefore, have a special place different from other academic or professional programs and departments in the university. For a long time policymakers and academics have sought to identify the space that teacher education should occupy between the school and the university. In many ways, this has been a persistent problem and never been adequately addressed by policymakers and university leaders. Just where should teacher education programs reside? Where do education schools fit between the needs of local schools, the expectations of the state (the preparation of an adequate supply of highly qualified teachers), and the academic demands and professional responsibilities to experiment and innovate? The case makes the point that such uncertainty stifles creativity and innovation.

The Louisiana study brings into sharp focus these tensions between the universities and the states. The concepts of faculty governance and academic freedom are challenged when states mandate fundamental programmatic changes and add courses and curricular expectations without consideration of the pace and routines of university governance systems. The deliberative stances of academic committees and the lengthy approval processes of academic governance are challenged by the needs of politicians and demands of educational reformers.

The Louisiana story is about how new expectations and demands for program accountability challenge the traditional prerogatives of colleges and universities. In this case, it is about the use of a new and ambitious method to measure program graduates' success in K-12 classrooms and attribute those successes to teacher preparation programs. Even though questions of validity and reliability were still being sorted out, state officials decided to use this methodology to judge the quality of teacher preparation programs. State decision makers declared their commitment to this process (and garnered widespread policymaker support for this approach), while universities were still grappling with its consequences. As Zeichner notes in his chapter, such accountability demands have far-reaching consequences for all university programs—with the possibility that other professional preparation programs will be held to similar expectations.

The authors of this case portray the challenges for education school deans caught between state policymakers and state officials and bureaucrats and the faculties and administrators in colleges and universities. Although we once might have talked about education deans having middle power and argued that being in the middle was a powerful position to be in, this case raises important questions about the ways to survive in what is now an seen as a highly contested arena. Viewed from the perspective of academic colleagues, those in the middle are bringing a message that demands compliance and conformity to external demands. For those at the state level who are engineering changes in schooling, those in the middle are seen as obstacles to change.

The case describes the reliance on prestigious commissions and ambitious reports that seems to be a consistent part of state policymaking. It describes the uses of those commissions and reports to set ambitious agendas (often failing to consult adequately with those who are the target for that change). It describes a process by which states use these commission reports to mandate programmatic changes, require national accreditation, rely on new or experimental performance assessments, and display publically the rankings and performance results of teacher preparation programs. It shows the ways that state power and authority are used to direct change for teacher education.

Elsewhere in this book, Zeichner raises questions about the use of value-added measures to assess teacher education programs. He highlights the fact that this is an imperfect means to determine the quality of teacher education. In this case, the authors suggest that the ambitions of policymakers have gotten far ahead of the reliability of this assessment technology. Although it remains an imperfect tool, the authors suggest, its use should be limited. The apparent unwillingness of policymakers to slow the pace of their accountability expectations or wait until they have assessment tools that are valid and reliable and can measure the efficacy of all teacher education programs is a challenge.

Discussion Questions

1. The case suggests that the position of colleges of education between K-12 schools and universities is uncertain, and that resolving this introduces ambiguities to core missions of colleges and universities. The authors of the case talk about the uncomfortable ground that they occupy as leaders of an education school. How does this contrast with the Montclair case (see Chapter 7) and the position that college occupies?

2. Zeichner suggests that value-added assessment does not lend itself to the kinds of determinations that Louisiana is trying to make, while Hess suggests that it is one of the few tools we have to make such judgments. Are policymakers likely to slow their adoption of value-added assessment measures until the technology catches up with their needs and expectations?
3. Does state intrusion into teacher preparation stifle innovation and experimentation? Does it affect research universities differently than it does comprehensive colleges and small liberal arts institutions?
4. Can value-added information for graduates of programs be useful in the absence of other data? Beyond gathering student score data from classrooms where recent program graduates are teaching, what other data and information should teacher preparation programs gather?

4

CHANGING STANDARDS, CHANGING NEEDS

The Gauntlet of Teacher Education Reform

Catherine Emihovich, Thomas Dana, Theresa Vernetson, and Elayne Colón

We contend that in few other states has the gauntlet of teacher education reform been thrown down more often than it has in Florida, the fourth largest state in the United States. In this chapter, we identify the teacher quality discourse and associated reform efforts that dominated Florida educational policies and practices during the period 2000–2010, which were closely linked to two pivotal events at the national level—the election of George W. Bush as President in 2000 and the passage of the landmark legislation No Child Left Behind (NCLB), in 2001. The prequel for Florida occurred in 1998, when Jeb Bush, brother to President Bush, was elected governor. Both Bushes proclaimed themselves as education reformers who believed strongly in the need to overhaul both public schools and schools of education for greater accountability, and both were firmly committed to neoconservative discourses that viewed educational attainment as a commodity necessary to ensure the United States maintained its economic edge in an industrialized global economy. The policy decisions in Florida that have unfolded over the last decade are a classic example of policymakers' desires to implement policies to improve teacher education programs through a patchwork of mandated rules and regulations, and their decisions underscore the importance of having strong evidence documenting the effectiveness of college-based programs. We argue that the lack of this evidence was one pivotal factor in the creation of alternative pathways and other regulatory policies that now threaten the continuing viability of college-based programs within the state.

Policymakers' frustration over their inability to improve educational outcomes often results in negative and counterproductive behaviors that undermine the very goals they seek to achieve. Throughout the first decade of the 21st century, especially in Florida, policymakers have become bullies to make their points about teacher quality. Although many in the halls of government in Tallahassee undoubtedly did not enter public service positions with the intent of becoming a bully, statute and rulemaking have clearly established a bully and victim relationship among policymakers and teacher educators. As is seen with children in bullying contexts (O'Connell, Pepler, & Craig, 1999), policymakers and teacher educators have fed off each other's perceived positions of authority and responsibility. Interactions between policymakers and teacher educators in approved teacher education programs can be described as strained as legislators give credence to biased reports decrying outdated modes of teacher preparation while teacher educators design professional development schools, job-embedded professional development programs, and other innovations that better connect the practical and theoretical. The divisive bullying dynamic continues with national personalities and think tank reports emboldening serial bullies who take advantage of a scant research/evidence base that could provide justification about the value of teacher educators' work.

The increasing clamor about the ineffective nature of educator preparation may be a direct comment on policymakers' frustration over their inability to change schooling through authoritative means such as mandating new curricula or requiring standardized end-of-course examinations. Plain and simple, it is easier to place blame for poor student performance on the teacher and the teacher's preparation program than it is to tackle the myriad of disconnected policies that constrain good schooling practices tied to whole school reform. This lack of connectivity impedes progress toward producing more equitable and socially desirable educational outcomes, especially for disadvantaged students (Levine, Belfield, Muennig, & Rouse, 2007; Portes, 2005). We believe that the need to improve educational outcomes and close the achievement gap is too important to allow the bullying metaphor to continue as a way of framing public policy debates. Fostering all students' learning certainly cannot be minimized given recent national reports which demonstrate that the United States lags behind most industrialized countries in achievement, particularly in mathematics and science (National Academy of Science, 2007). In addition, the continuing and persistent achievement gap (or, to use Gloria Ladson-Billings' [2006] more pithy descriptor, "education debt") between white students and students of color is not just unsustainable for social justice reasons, but it also presages a grim economic future as minority students from low-income families become the new majority in U.S. public schools but who still lack access to equal opportunities for learning.[1] We argue, however, that subjecting teacher education programs to constant demands for meeting new standards, and adopting new preparation routes in reaction to rapidly changing state and national policies, leaves battle scars that time alone may not heal. Weary teacher educators often feel that working in this field constantly involves picking up the gauntlet thrown down by legislators and/or policymakers as each new ambitious teacher education reform is launched as a way of improving educational outcomes for America's children (Imig & Imig, 2008). Our experiences serve as a cautionary tale that underscores the challenges and difficulties teacher education programs are likely to face as state budgets continue to decline, and the demands for greater educational accountability continue to be pressed at state and national levels.

The Case in Florida

For many years, policymakers in Florida and elsewhere claimed that highly qualified educators are not available to all students. They promoted a vision of a functional education system built on world-class curriculum standards with measurable increases in levels of student achievement. As is the case in large states, it's a daunting task to move a large and complex public education system with 2.6 million students, 170,000 teachers, and 3,600 schools in 67 consolidated school districts (Florida Department of Education, n.d.).

Despite the challenges, a laser-like focus on teacher quality emerged. One theme throughout the past ten years in Florida has been on teacher recruitment and the role university-based teacher education played in meeting teacher workforce needs. From 2003, when 22,000 new teachers were hired, to 2008, with 11,000 new teachers entering classrooms, the percentage of new teachers prepared by the state's public colleges and universities varied slightly but never accounted for more than 45% of new teachers. Of the 15,707 individuals who received an Initial Professional Certificate during 2006–2007, approximately 35% ($n = 5,400$) completed an approved Florida teacher education program, 35% presented an out-of-state certificate for reciprocity, 16% completed a course-by-course transcript analysis, and 24% completed an alternative certification route (Hebda & Pfeiffer, 2009).

We begin by describing the background of the state program approval process for teacher preparation programs to provide a context for changes that have taken place over the last decade. Florida's government—both the executive and legislative branches—has long played a major role in regulating the preparation of educators. Either through statutory authority or through regulatory means, the preparation and certification of teachers and other professional educators lies firmly within the government's purview. In 1996, Governor Lawton Chiles convened a Governor's Task Force on Education to "develop consensus about meeting Florida's education challenges for the next decade and to assist in accomplishing that task" (Governor's Commission on Education, 1996–1998). One of the six areas pursued in achieving this goal was "to recommend appropriate reforms in professionalization efforts, teacher training and retraining, increased diversity of the education workforce, collective bargaining practices, staffing patterns, and other workforce changes that could enhance student learning" (Governor's Commission on Education, 1996–1998). This task force established the context for policymakers to mandate changes in teacher education programs to improve student learning, a context that became progressively more politicized as the political dynamics in Tallahassee shifted over time.

The election of Jeb Bush as Florida's governor in 1998 marked the beginning of fundamental changes in the teacher education landscape as he began to exert tighter reins on educational reform in the state. Prior to 2002, the development of standards for state approval of Florida teacher preparation programs was largely in the hands of the Education Standards Commission (ESC), an appointed body of 24 practicing educators and citizens who were nominated by the Commissioner of Education, appointed by the State Board of Education, and confirmed by the Senate. Members served for three-year staggered terms. The membership included twelve teachers, one professional development director, one school district personnel officer, one principal, one school district superintendent, three university representatives (including two deans of education from public institutions and one from a private institution), one community college administrative representative, and four citizens to include two current school board members and two lay citizens.

Among other responsibilities, the ESC members were responsible for recommending to the State Board of Education preservice teacher education program standards, as well as standards to measure the competence of Florida teachers. Title XVI, Chapter 240(2) of the Florida Statutes from 1997, states:

> A system developed by the Department of Education in collaboration with institutions of higher education shall assist departments and colleges of education in the restructuring of their programs to meet the need for producing quality teachers now and in the future. The system must be designed to assist teacher educators in conceptualizing, developing, implementing, and evaluating programs that meet state-adopted standards. The Education Standards Commission has primary responsibility for recommending these standards to the State Board of Education for adoption. These standards shall emphasize quality indicators drawn from research, professional literature, recognized guidelines, Florida essential teaching competencies and educator-accomplished practices, effective classroom practices, and the outcomes of the state system of school improvement and education accountability as well as performance measures.

Initial state program approval also called for a process that:

> incorporates those provisions and requirements necessary for recognition by the National Council for Accreditation of Teacher Education and provides for joint accreditation and program approval by the state and the National Council for Accreditation of Teacher Education for those units seeking initial or continuing accreditation.

For initial program approval an institution responded to 19 standards, whereas those institutions with full approval responded to five continuing approval standards and each institution's programs were reviewed every five years. In essence, a state approved program also had to be NCATE-accredited in order to have its graduates qualify for full licensure upon graduation without needing to meet additional requirements.

State Board of Education Rule 6A-5.066, the teacher preparation approval regulation created in 1998 to implement the statutory legislation, included additional requirements for state program approval. Those requirements addressed admission standards (minimum GPA, 40th percentile score on a nationally standardized college entrance examination, or, in the case of a graduate program, completion of a bachelor's degree from an accredited institution), graduation standards (a minimum GPA and an overall passing rate of all program completers at 80% on the written examination). Institutions seeking continuing program approval were required to use graduates' satisfaction, responsiveness to local school district needs, an annual review involving primary stakeholders, and the teaching of higher level mathematics instruction at the appropriate grade level.

In addition to establishing standards for program approval, the Education Standards Commission was assigned the responsibility for developing a set of standards for effective instructional practice for all educators. In 1998, the ESC presented to the State Board of Education for its approval the Florida Educator Accomplished Practices (FEAPs): 12 performance-based, essential practices of effective teaching at the pre-professional, professional, and accomplished levels. Definitions for the accomplished practices and suggested indicators at each level were recommended. The 12 FEAPs included assessment, communication, continuous improvement, critical thinking, diversity, ethics, human development and learning, knowledge of subject matter, learning environment, planning, role of the teacher, and technology. The definitions and suggested indicators are found in State Board of Education Rule 6A-5.066. School district educator evaluation plans included the FEAPs as they became known, and educator preparation programs were required to include demonstration of the FEAPs at the pre-professional level as a requirement for program completion of a state approved program.

In 1999, following the inauguration of Governor Jeb Bush, another task force, the Commissioner's Task Force on Teacher Preparation and Certification, was authorized in FS 240.529(1). Appointed by Commissioner of Education Tom Gallagher, this second task force[2] comprised of representation from presidents of public and private colleges and universities, deans of education, community college presidents, school superintendents and high performing teachers, recommended to and received approval from the State Board of Education in August 2000 to implement a required core curriculum for all teacher preparation programs. The core curriculum included specific, prescribed semester hour courses in general education (Table 4.1) and professional education (Table 4.2).

At first reading, it seems that these mandated curriculum requirements would be favored by teacher educators, since they included a heavy dose of professional education courses. In reality, the implementation of the core curriculum requirements

TABLE 4.1 General Education Course Requirements

Requirements	Credit Hours
English to include writing, literature, and speech	9 semester hours
Science to include earth science, life science, and physical science, with a minimum of one associate laboratory	9 semester hours
Math to include college algebra or above and geometry	9 semester hours
Social science to include American History and general psychology	12 semester hours
Humanities, philosophy, and fine arts	6 semester hours
Total General Education Required Hours	**45**

TABLE 4.2 Professional Education Course Requirements

Requirements	Credit Hours
Reading-literacy acquisition for PreK-primary and elementary education teachers as well as teachers who were the primary teachers of English, which included special education and English/language arts teachers	12 semester hours
Teaching reading for all other content areas	3 semester hours
Classroom management, school safety, professional ethics, and educational law	3 semester hours
Human development and learning	3 semester hours
Assessment to include understanding of the content measured by state achievement test, reading and interpreting data, and using data to improve student achievement	3 semester hours
Total Professional Education Course Required Hours	**12–21**

from the Gallagher Task Force into State Board Rule created a conflict between the State University System rules for maintaining 120-credit-hour bachelor's degree programs and the teacher educator community. Because of the prescriptive nature of the 45 semester-hours of general education curriculum within the first 60 hours of a candidate's 120-hour program, and because an additional 9 semester-hours were required by the articulation agreement with the community colleges, in some cases candidates had a total of only 6 hours of elective courses available during their lower division coursework. In other cases, such as music education, in order for those candidates to complete the required music performance hours and methods courses in choral and instrumental music, their programs far exceeded the required 120 credit hours. Finally, students who chose an education major late in their undergraduate career, or who wanted to transfer into education from an out-of-state university, found it almost impossible to meet all these requirements in a timely fashion. In short, any flexibility in program development and implementation was removed by the

Task Force recommendations. We will describe later how this recommendation thrust the college programs into a double bind situation.

Other statutory changes that were enacted in various education bills in 1999 included the removal of cut-off scores for nationally standardized college entrance exams for admission to teacher education, but they were replaced with a requirement which amended SBE Rule 6A-5.066 in 2000, and became effective in the 2000–2001 academic year. This requirement stipulated "mastery of general knowledge including the ability to read, write, and compute by passing the College Level Academic Skills Test, or the National Teachers Examination, or a similar test approved by the State Board of Education." Also included in FS 240.56 was statutory language for school district supervisors of student teachers to include successful demonstration of management strategies that "consistently result in improved student performance," demonstration of effective classroom management strategies and incorporation of technology into classroom instruction, as well as linking instructional plans to the Sunshine State Standards (Florida's K-12 subject area student learning standards). In addition, field placements for teacher preparation students were to represent the full spectrum of school communities, including, but not limited to, schools located in urban settings. Appearing for the first time in 2000, attention to English Language Learners (ELLs) was included in the public accountability and state approval for teacher preparation programs legislation. State approved teacher preparation programs had to incorporate English for Speakers of Other Languages (ESOL) instruction so that program graduates would have completed the requirements for teaching limited English proficient students in Florida public schools. Language to ensure attention to the Florida ESOL requirements was added to SBE Rule 6A-5.066 at the same time. We emphasize again the point that, although these requirements appear reasonable (e.g., strong teacher education programs should include strategies for teaching ESOL students), they increased the number of required hours in an undergraduate degree program, and made it virtually impossible for students to stay within the 120-hour mandate without removing non-mandated requirements, such as subject-matter courses beyond the general education requirements for elementary education majors.

The establishment of very specific and restrictive requirements signaled the beginning of limiting the ability of teacher preparation institutions to respond to the growing demand for alternatives to traditional routes into licensure at the same time that the state was experiencing an increased shortage of qualified teachers. The limitations were twofold in nature. First, complying with all these requirements placed an enormous financial burden on institutions in employing sufficient faculty to teach all the mandated courses. Just as one example, meeting all the ESOL requirements in all educator preparation programs requires having three to five faculty members who are experts in this area. Institutions that employ extensive numbers of adjuncts to reduce the cost then face problems in meeting NCATE standards for faculty/student ratios. Without additional state support, institutions simply did not have the resources available to maintain a traditional program and develop alternative programs. Second, state institutions were unable to petition for relaxation of these requirements since the State Legislature strongly believed statutory oversight and regulation was necessary in state funded programs. As Kumashiro noted:

> At the same time that policy makers ask teacher preparation programs in higher education to do more to prepare teachers for certification, they authorize and fund fast-track programs that, by definition, do less. This contradiction functions not to raise the quality of teachers, but to undermine teacher preparation in higher education altogether, since preservice teachers can only receive less, not more preparation from both the fast-track and the "traditional" programs that must now divert enormous resources to documenting that they meet the increasing requirements and restrictions.
>
> (2010, p. 60)

In effect, teacher preparation programs in colleges of education were held accountable for meeting mandated requirements that they did not fully support, but at the same time were severely criticized for not being flexible or creative to provide alternatives to a growing number of adults who did not fit the profile of the typical undergraduate student majoring in teacher education. In our judgment, this scenario fits the profile of a classic double bind situation.

The Rise of Alternative Certification Pathways: 2002–2006

A pivotal event occurred in 2002 with the passage of the class size reduction amendment to the Florida Constitution that limited the number of students at various grade levels, and outlined serious penalties for schools that failed to adhere to them. At the same time, Florida experienced a huge growth spurt from 2002 to 2006 with annual increases of about 418,000 people (Smith & Cody, 2010). Although Florida remained a destination for retirees, those 65 and older accounted for only 15% of migration to the state. The largest group moving to Florida during 2002–2006 was younger people in their twenties and thirties, often with children in tow (Smith & Cody, 2010). Both legislators and state education officials became increasingly concerned that Florida would not have enough teachers to meet the demand, especially since the state universities never produced more than 45% of the state's teaching needs, even under the best circumstances. For this reason, they authorized initiatives to create flexible and short-term ways for career changes to get into classrooms. In this section, we describe four alternatives that were quickly put into place: district alternative certification programs; pathways tied to subject area examinations and the test developed by the American Board for the Certification of Teacher Excellence (ABCTE); Educator Preparation Institutes; and Professional Training Options.

District Alternative Certification Programs

In 2002, a District Alternative Certification Program (DACP) was approved by the Legislature in FS 1012.56. A school district could choose to adopt the state developed alternative certification model or a district developed alternative certification program approved by the Florida Department of Education. The district alternative certification model requires that individuals who are enrolled be temporarily certified in a subject area prior to entry into the DACP. Temporary certification is based on a subject-matter degree held at the bachelor's or higher degree, a bachelor's degree with a 2.5 GPA in the content area, or a bachelor's degree from an accredited institution and a passing score on the Florida Teacher Certification Subject Area Exam (SAE) in the subject area. The first year of teaching serves as the substitution for the traditional student teaching internship. This route has proven to be extremely popular, since over 1,600 people have completed DACPs as of 2008–2009.

ABCTE and Subject Area Examinations

A second route for temporary certification is a passing score on the examination given by the American Board for Certification of Teacher Excellence (ABCTE). ABCTE was formed in 2001 from a grant from the United States Department of Education to develop alternative certification routes for baccalaureate degree holders seeking to change careers into the field of education. ABCTE did not require coursework or other requirements typically found in a university-based teacher education program. Instead, readiness for certification was determined by successful completion of a set of examinations (American Board for Certification of Teacher Excellence, n.d.). Florida began to accept ACBTE certification in June 2004. As noted by Glazerman, Seif, and Baxter (2008), Florida had the greatest growth in this certification route of any state. As of the 2007–2008 school year, 94 individuals who achieved certification through the ABCTE examination from its inception were teaching in Florida public and private schools (Tuttle, Anderson, & Glazerman, 2009). Individuals with a bachelor's degree could also be issued a temporary license if they received a passing score on the Florida Subject Area Examination and passed a criminal background check.

Educator Preparation Institutes

The Educator Preparation Institute (EPI) is most frequently offered at community colleges, but it can also be offered at a public or private university to provide an alternative route to teacher certification for mid-career professionals and college graduates who were not education majors. To encourage both community colleges and state universities to develop these programs, in 2004 the Florida Department of Education created a grant program known as SUCCEED Florida. Although not required to do so in order to create an EPI, institutions could apply for funds to develop pilot programs that were expected to pay for themselves through tuition revenue once the funding period ended. For the first time, a state supported college or university could enter the alternative certification market while still maintaining a traditional teacher education program. As of June 2010, 30 Florida community college or universities offered EPIs.

An examination of EPIs across the state demonstrated there are remarkable similarities. Generally six to eight months long, many programs require six to seven courses and two field experiences. Courses tend to be blended, requiring evening or one full-day Saturday meeting per month (8 a.m.–4 p.m.) and completion of an online component. Some EPIs are fully online. One example is the EPI at Pasco-Hernando Community College. The curriculum of this EPI program consists of seven courses relating to the standards of the Florida Educator Accomplished Practices (FEAPs), and two field experiences (diversity and professional foundations—totaling ten days). The courses comprise segments that provide students with the skills set and knowledge base necessary for those in the teaching profession. Each course is from four to six weeks in length and is presented either in a classroom setting or online. Students are also required to create a portfolio that documents their competency of the FEAPs and students must also pass all required sections of the Florida Teacher Certification Exam. The classroom model combines face-to-face and online learning and can be completed in as little as seven months. The online model can be completed in two semesters. The model provides for minimal supervised clinical preparation, and the student teaching component is primarily addressed by counting the teacher's first year of teaching as part of their internship.

In response to growing competition from the community colleges' EPI models, and to demonstrate flexibility in offering multiple pathways to teaching, the University of Florida (UF) and several other traditional four-year public universities decided to pursue the less restrictive EPI alternative certification model. In strong contrast to the Pasco-Hernando example above, the participants in the University of Florida program complete one full academic year in a classroom with a mentor teacher while also completing 36 semester hours of coursework toward the master's degree. The program begins in the summer semester and ends the following spring with graduation from the program. Although the universities adopted the EPI model, a strong clinical component was retained, based on the belief that clinical preparation was essential in any teacher education program even in a time limited model (AACTE, 2010; Lampert, 2010).

Professional Training Option

A fourth alternative, the Professional Training Option (PTO), refers to an opportunity for students who are enrolled in one of Florida's four-year, degree-granting colleges or universities to complete 15 semester-hours of professional education coursework required for Florida certification. This can be accomplished as an education minor, or through another approved route at the institution. Graduates who complete the PTO have a statement added to their transcript signifying that they have completed the PTO and are therefore exempt from the requirement for additional professional education coursework in order to be issued a regular teaching certificate in Florida. Although eligible for a temporary certificate at the completion of their baccalaureate degree, the graduate has only to complete one year of successful teaching in a Florida school district, and receive passing scores on the Florida Teacher Certification Examination in order to qualify for a professional certificate. In contrast to EPIs, which require the demonstration of the 12 Florida Educator Accomplished Practices (FEAPs), the PTO requires only that participants are exposed to the FEAPs. Although the EPI requires actual supervised clinical practice, the PTO requires only observation in a K-12 classroom.

These new pathways presented traditional, college-based programs with multiple challenges. First, the removal of any mention in statutory language of aligning state program approval standards with the National Council for Accreditation of Teacher Education (NCATE) paved the way for school districts and online providers to develop programs that would qualify for state approval only under more relaxed standards for alternative pathways, since in many cases these programs would have been unable to meet the more stringent standards expected for NCATE approval.[3] We note, however, that these NCATE accredited institutions face an inherent conflict of interest in maintaining their traditional, state approved program (which is necessary for continued state funding), and having alternative routes which are not NCATE accredited. One resolution to this conflict may well be that the state universities will choose to drop NCATE accreditation for their traditional programs, since there is no competitive advantage in offering a nationally accredited program if the alternative certification programs only need state approval for licensure.[4]

The second challenge is the issue of who has authority over teacher education programs. In her Vice-Presidential Address for Division K at the AERA national conference in 2004, Pam Grossman (2008) framed the issue of teacher regulation as a jurisdictional challenge to university-based teacher education programs, and argued that teachers and college faculty were in serious danger of losing control over two key professional tasks: the preparation of new professionals and the production of academic knowledge for the profession. The critical issue is not whether the state has the power to regulate the profession. The primary concern for teacher educators is the set of assumptions used to guide decisions about who can be licensed and under what conditions, and about the nature of teaching and its status as a recognized profession. Grossman used the development of the ABCTE exam to press her point about the lack of a strong evidentiary base to defend college-based, teacher education programs:

> The designers of ABCTE have, in effect, co-opted the process and even the language of standards-based movements of subject matter organizations,

such as NCATE and NBPTS, to develop their approach. We cannot, thus, disagree, in principle with how they reached consensus on what teachers need to know and be able to do or how they developed an assessment plan aligned with those standards. Where we need to disagree—and to be able to mount a strong critique—concerns their claims of what it takes both to teach and to prepare teachers. Although some of that argument will always be driven by normative assumptions, professional judgment, and ethical commitments, our critique would be substantially enhanced as well as more professionally responsible and publicly credible, if we had a sound research base with which to refute their claims or support our own.

(Grossman, 2008, p. 13)

Grossman's call to teacher education researchers to meet this challenge has not gone entirely unheeded. Since her address in 2004, several studies have documented with varying success the effectiveness of college-based preparation programs on student learning, including her own research conducted for the Carnegie Foundation on Teaching (Boyd et al., 2008; Darling-Hammond, 2006). However, the research base is admittedly thin, complicated by the fact that few rigorous research studies exist which document teachers' impact on student learning regardless of the teacher preparation route they chose. In the cases that do exist, most notably the research conducted by Noell and Burns (2006) on Louisiana's teacher education programs, the results do not necessarily demonstrate that traditional programs produce higher student learning gains on standardized tests than do the alternative programs.

By 2006, the scope of Florida's teacher preparation programs included a variety of pathways into initial certification. Although these alternative pathways certainly provided the easier access to becoming a teacher that the policymakers demanded, the potential downside is that students in low-income schools are far more likely to have teachers prepared without a strong grounding in clinical preparation, which may impact their future achievement (Peske & Haycock, 2006).

The Second Wave of Reform: 2006–2010

In 2006, a new governor was elected, Governor Charlie Crist, and like many of his predecessors sought to put his own stamp on state education policies even though the underlying political dynamic remained unchanged, since he was still a member of the Republican Party. After Crist was elected, issues of teacher quality continued to capture the attention of policymakers. A State Board of Education Workshop held in October 2006 focused on teacher preparation, teacher recruitment and retention, and teacher compensation. The workshop was led by the K-12 Chancellor, Cheri Pierson Yecke, for the purpose of assessing how well Florida was at acquiring and keeping (a) enough teachers with (b) the level of effectiveness necessary to ensure high levels of student achievement. Yecke's comments signaled further dismissal of accredited teacher preparation programs as a primary source of new teachers. There was a growing acceptance among Florida policymakers that multiple pathways into teaching were necessary, and that the defining feature of teacher quality was dependent on high levels of student achievement, and not on the type of certification pathway candidates followed to enter teaching. We want to stress the point that the state university education deans are certainly not averse to the idea of assessing teacher education program quality based on student learning gains, nor are they opposed to having multiple pathways, but they are also deeply concerned that many teachers are entering the profession without the required reading and ESOL training, and classroom management strategies mandated for the college-based programs for state approval.

In considering the state of teacher quality in October 2006, the State Board of Education reviewed findings and recommendations from selected national research studies. The national studies examined offered a set of common themes that had now become depressingly familiar: the regulatory approach that requires teachers to complete formal training is flawed; teacher education programs are of poor quality and do not adequately prepare teachers for contemporary classrooms; alternative pathways into teaching are necessary; approaches to certification should be streamlined; and, teacher evaluation and compensation should be determined by students demonstrating satisfactory achievement gains. Table 4.3 provides a summary of the reports considered by the Board.

One of the conclusions drawn from the State Board workshop was that efforts to simplify routes into classrooms needed to be fast-tracked to meet teacher supply issues. The state education deans clearly understood the point that, because historically never more than 45% of the state's teachers had ever been produced at their institutions, alternative pathways were necessary, but they contended that these routes had to be held accountable to the same preparation standards as the college-based programs, which was certainly not the case. From the policymakers' perspective, the underlying assumption was that by expanding the pool of candidates school officials would have a choice of highly capable teachers and, perhaps, at some point would have sufficient candidates to replace teachers who met entry qualifications but whom they believed were not doing an adequate job of increasing student learning gains. At the time this workshop took place (2006), no hard evidence was available to support their beliefs about teacher performance, and the state university system deans strongly felt their programs were unfairly being characterized as producing unqualified teachers.

In October 2006, the Florida State Board of Education engaged in a revision of Rule 6A-5.066, which as we noted earlier in this chapter governed the teacher education program state approval process. The team of collaborators from multiple teacher preparation institutions who formed the revision committee were tasked by the Florida Department of Education to envision a new, all-encompassing teacher preparation rule that would incorporate all new and existing avenues into teaching while lifting restrictions that hampered innovation in the past. The committee

TABLE 4.3 Summary of Reports Considered by Florida State Board of Education, October 2006

Report	Summary
Reforming Education in Florida (Peterson, 2006)	The assessment reported in this document from the Hoover Institution was undertaken at the invitation of Governor Jeb Bush and then Florida State Board of Education chairperson Philip Handy. This assessment identified multiple education-related reforms (accountability, world-class curricula, teachers, school choice, Pre-K, and Class Size and Resource Utilization) initiated under Governor Bush and offered 30 major policy recommendations, including: (1) State-endorsed alternative certification routes through school districts and via Educator Preparation Institutes (EPIs) based in community colleges as well as four-year colleges have worked well in meeting the challenge of recruiting new teachers and should be continued. (2) Florida should adopt a streamlined approach to certification that allows principals to hire anyone with a bachelor's degree who can demonstrate substantive knowledge (via an exam) and pass a background check. Specific coursework or other preparation should not be required. (3) Teachers with students who do not make satisfactory achievement gains should be dismissed. (4) Florida should continue to invest in innovative teacher compensation models, including the STAR (Special Teachers Are Rewarded) system and allocate funds to schools based on gains in student performance. The task force was chaired by Chester E. Finn as part of the work of the Hoover Institution.
Identifying Effective Teachers Using Performance on the Job (Gordon, Kane, & Staiger, 2006)	The report sponsored by the Brookings Institution criticized an approach to improving the quality of teachers by raising standards for entry into the profession citing research that suggests initial teacher credentials "have little predictive power in identifying effective teachers" (p. 2). The report also promoted revamped teacher evaluation systems that include student achievement impact, and opening the classroom door to teachers who lack traditional certification but can demonstrate success with students.
Educating School Teachers (Levine, 2006)	The Levine report lambasts educator preparation and concludes that a majority of teachers are prepared in university-based programs with low admission and graduation standards and with the least accomplished professors (pp. 26–28). Several teacher education programs are cited as models, but most are characterized as having faculty, curricula, and research that are outdated and disconnected from school practice.

worked hard to find ways to reduce the number of standards and the lengthy processes associated with meeting those standards at the initial and continuing approval level. Three standards were finally agreed upon—Core Curriculum, Candidate Performance, and Continuous Improvement (see Appendix A). After considerable discussion, along with arguments that at times occurred, the final decision to use three standards was agreeable to all teacher education providers from both community colleges and the state universities.

With the move from specific, mandated courses to competency demonstration through authentic performance-based tasks, the rigorous expectations for systems of accountability emerged. To track the mastery of certain skills and competencies, institutions employed complex electronic tracking systems to document the progress of educator preparation candidates. In some cases, institutions used commercially available software packages (e.g., TaskStream), and in other cases institutions developed their own system. As an example of a university-based system, the University of Florida developed its electronic system after the 2003 NCATE/FL DOE accreditation review. This decision was based in part on the fact that the university employed a data management system that did not readily interact with the commercial systems available at the time. In addition, commercial systems that track candidate progress often require students to purchase an account, which creates a financial burden, or place the onus on students to "populate themselves" in courses or programs. In contrast, the UF system was designed to pull course rosters and other pertinent student data (e.g., admissions information) from the university systems and link this information to performance evaluations in courses and field experiences. This helped to ensure the accuracy of the data and allowed users to control the maintenance and upgrades to the system. Furthermore, the system that was developed is dynamic; as standards are revised or added, the system easily accommodates these changes. To keep pace with the changes made to the standards and rules for educator preparation requires enormous resources. There are approximately seven faculty and staff members and a handful of graduate students that devote a significant portion of their work time each week to maintaining or continuously improving this system in order to meet the demands of state and national reporting requirements, at a cost of over $200,000 per year.

New Threats to College-Based Teacher Education Programs

By the beginning of the 2007–2008 academic year, Florida deans of education felt that the rapid-fire changes experienced over the last seven years were stabilizing and that relations with the FL DOE gradually were improving. Many institutions offered alternative programs in the form of the Educator Preparation Institutes (EPIs) for post-BA students, or the Professional Training Option (PTO) by offering an education minor that enabled undergraduate students to complete the requirements for temporary licensure. At that point, only ten community colleges had participated in the pilot project that was begun in 2004 to develop their own teacher education programs through the Educator Preparation Institutes, and several State University System (SUS) institutions had developed strong partnership programs with their local community colleges in a 2 plus 2 transfer model. As one example, the University of Central Florida set up branch campus programs within several local community colleges to facilitate transfer into their program at the junior level. In contrast, at the University of Florida, the official university policy was to reduce the number of transfers for any major from the community colleges into UF, since the president wanted to reduce the size of the undergraduate footprint in order to concentrate more on graduate education. This policy created a problem for the University of Florida teacher education programs at the elementary and early childhood levels, since 40–50% of applicants were transfers from community colleges, primarily from Santa Fe Community College (now Santa Fe College). Despite this transfer reduction, Santa Fe indicated they were still not interested in developing their own program, since their particular market niche in the community college world was based on their strong affiliation with the University of Florida.

In January 2008, under pressure from the community college presidents, and given the favorable reaction to the EPI program, legislation was passed that authorized all community colleges the ability to offer four-year teacher education programs in addition to the EPIs they had already developed. By June 2008, 17 of the 28 community colleges had now developed at least one program. With this action, the stakes for maintaining enrollment targets were considerably raised for the SUS institutions, since tuition costs were significantly less at the community colleges (see Table 4.4), and with no hard evidence to assert the claim that enrolling in an SUS institution would produce a better teacher the SUS schools were likely to lose their market share.

Another threat the state university programs faced was the increasing pressure to link the performance of teacher education graduates with student learning outcomes on the Florida Comprehensive Assessment Test (FCAT). Over the last few years (2003–2007), the FL DOE had refined its capability to match individual teachers with their students' scores in a given year along with the institution where they were prepared. In June 2009, the Commissioner of Education, Dr. Eric Smith, summoned deans/directors of all teacher education programs (including community colleges and independent colleges) to a meeting in Tampa, where for the first time they received data from the FL DOE on student FCAT achievement gains for 2008–2009 that were tied to the identity of each classroom teacher who completed their program in 2007–2008, and who taught each course

TABLE 4.4 Comparison of Credit Hour Costs by Institution

Type of Institution	Average AA Cost Per Credit Hour	Average Baccalaureate Cost Per Credit Hour
State University	—	$140.64
State College	$86.77	$94.63
Community College	$86.69	—

in reading/language arts and math in Grades 4 through 10 in public schools during the 2008–2009 school year. Furthermore, the data compared all SUS institutions with independent colleges, private providers, community colleges, and district-based alternative certification programs. These data were handed out in the meeting in sealed envelopes, and as each dean opened her/his envelope the room became completely silent. As one glanced around the room, it was clear from the sagging shoulders and crestfallen faces that some programs, including a few in the SUS system, did not fare well in this comparison. Afterwards, the SUS deans met privately and discussed a plan of action, since they realized that if these data were publicly reported by the media it could be used to undermine even further the college-based programs.[5] Even if all programs performed well, that finding would not have been sufficient unless the deans could demonstrate that the graduates of college-based programs did significantly better than those who were prepared in other programs. The fact that some of the community colleges' graduates and district-based alternative certification programs performed as well as the SUS graduates led to the inescapable conclusion that even more students would be attracted to these programs. This fact did not escape the attention of the Florida Legislature, which began drafting legislation in December 2009 to authorize a study of the cost-effectiveness of all state teacher education programs.

A concern for deans of education was that the data analysis done by the FL DOE was overly simplistic and misleading. In effect, they simply matched the scores of each teacher with the institutions where they were prepared, and then reported the percentage of teachers that either had 50% of their students maintain academic performance at level 3 (proficiency), or whose students increased in achievement by at least one level. No attempt was made to differentiate the type of schools where teachers taught, or to control for other impact variables such as teacher characteristics, student demographics, family income, previous student performance, and school climate effects. Furthermore, because the FCAT is given in March of each year, teachers were assessed on gains (or losses) in students' performance based on their performance on the FCAT from the previous year. In other words, students took the FCAT in 4th Grade with one teacher and then took it again in 5th Grade with a different teacher. A better measure of a teacher's performance is to consider how his/her students performed from the beginning of the school year in September to their performance at the end in June. Additional concerns were that during the two test dates the pool of students could have dramatically shifted given the high mobility rates in many urban schools, and the lack of instructional opportunities over the summer due to budget cutbacks meant many students might have fallen behind without any teacher intervention. Several deans were aware of the value-added research done by George Noell (Noell & Burns, 2006) at Louisiana State University, and it was suggested that he be invited to give a presentation at one of the state-wide meetings of the Florida Association for Colleges of Teacher Education, a state affiliate of the American Association of Colleges for Teacher Education, to illustrate the value of a more complex analysis to the FL DOE. The deans fully recognized that this more complex analysis could result in demonstrating that the SUS institutions were still not performing significantly better, but this was a political risk that had to be taken. Dr. Noell presented his research at the October 2009 FACTE meeting, and as of this writing the FL DOE is considering using his approach, although no formal decision has been reached. The SUS deans also considered funding their own study, but decided against doing so, since there were no resources available to support it, and they felt it would be perceived as a biased report if it originated under their auspices.

But none of the events that have already been discussed prepared the education deans for the magnitude of the assault on teacher education (and public education) that took place during the spring 2010 Florida legislative session. In January 2010, Jon Rogers, the legislative liaison to the Florida Board of Governors (an organization that oversees the regulation of all state universities), warned the deans to track the passage of a new bill rapidly moving through the Legislature. This bill, which was numbered Senate Bill 6, was sponsored by Senator John Thrasher. The bill was intended to be a comprehensive education personnel initiative that targeted K-12 school districts and charter schools' personnel and assessment policies, and mandated new requirements for state approved educator preparation programs and professional certification. A companion bill in the House, House Bill 7189, was also quickly filed that was identical to Senate Bill 6, in order to eliminate the possibility of having any differences in each bill taken up in committees to resolve the discrepancies.[6] In short, both bills were on a very streamlined timetable for quick passage and the Republican House and Senate members had their marching orders[7] not to propose any amendments that could slow the process down. It was widely speculated that these bills were under the sponsorship of former governor Jeb Bush, and as some experienced political observers commented the intent was to break the teachers' unions in Florida.[8] The legislation would have locked in place many of the education reforms Bush had championed during his time as governor. The provisions of Senate Bill 6, if enacted, would have had a far-reaching impact on policies and practices for both schools and institutions of higher education. Appendix B describes some of the more significant provisions, based on an analysis by the staff of the Education PreK-12 Committee in the Florida Senate (2010) that outlined the changes to previous statutory amendments.

To say that these provisions, once publicly known, created immediate concerns among all education deans/directors in Florida would be a huge understatement. Essentially, Senate Bill 6 was not only focused on ending tenure for K-12 teachers and school-based administrators; it also compromised the viability of teacher preparation programs in colleges and universities. This point was evident in the provision for eliminating the use of "degrees held" in setting salary schedules (which would have seriously undermined the continuation of master's and doctoral programs); in the provision tying continued program approval to measures of teacher effectiveness that were poorly defined and not validated by research; in the provision on eliminating the 10% admission waivers (which, contrary to legislative opinion, were used primarily for students who changed majors and did not meet all the mandated admission requirements, and not for "less qualified students"); in the provision requiring additional field experiences with placements with highly effective teachers despite no means for assessing their effectiveness and no incentives to teachers and their schools for taking on more responsibilities; and, finally, in the provision that stipulated all educator preparation programs would be continually subjected to heavy scrutiny by Office of Public Policy Analysis for Government Accountability (OPPAGA). In addition, despite repeated and insistent attempts by the education deans to obtain clear reassurances that all provisions in Senate Bill 6 would apply *equally* to any provider in the state who engaged in educator preparation, the language in the full bill was rather slippery on this point, since it continually referenced state approved education programs in postsecondary institutions. Our concern was that the double bind pattern regarding overregulation of college-based programs and lesser accountability for alternative providers would resurface and accelerate the decline in the number of teachers prepared through college-based programs.

What ultimately saved the colleges' programs was not the furious lobbying all deans engaged in, but, as the political world knows, Senate Bill 6 was vetoed by Governor Charlie Crist, who received over 100,000 calls and emails from teachers and administrators across the state, as well as from many citizens who were outraged at the blatant attempt to eliminate teachers' tenure and to control education policies at the local level. Even with this massive outpouring of resistance, had Governor Crist not been seriously behind in the primary race for the Republican nomination for the Senate against his opponent, Marco Rubio, the former Speaker of the Florida House, he probably would not have vetoed it. In doing so, he sealed his own political fate by deciding to leave the Republican Party and run as an independent candidate in the Senate race. Crist ultimately lost his Senate bid by a wide margin.

Florida's Future in Teacher Education

The case of teacher education preparation in Florida is instructive not just because its large size and growing diversity makes it a good harbinger of trends to come, but also because the education reforms that have either taken place or were proposed over the last ten years have been adopted in some form or another by other states. Without any doubt, a revised version of Senate Bill 6 will likely emerge in future Florida legislative sessions. Almost immediately after Governor Crist vetoed the bill, the Florida Board of Education issued a statement expressing their disappointment with his action, and called for a review of the provisions that produced the most vocal opposition. Even if some form of tenure is preserved, the idea of linking teacher compensation to student achievement is now firmly established, and some variant of this proposal will resurface. In the absence of any hard evidence attesting to value of advanced degrees, all the SUS education deans fully expect this idea to be revisited as well. The recent media release of a study conducted by two Harvard researchers (Chingos & Peterson, 2010), who co-incidentally are members of Harvard's Program on Education Policy and Governance, and whose advisory committee is chaired by Jeb Bush, suggests that the former governor still has a strong influence over educational policy in Florida. This report, which asserts that master's and doctoral degrees have no impact on teachers' effectiveness for increasing student achievement, based on using Florida's FCAT data from 2002–2008, will almost certainly be used to bolster this argument when the issue resurfaces in the Florida Legislature in spring 2011. The fact that the Chingos and Peterson (2010) study has not been vetted through rigorous peer review in a reputable journal is not likely to negate its value in determining public policy regarding the value of advanced degrees from colleges of education.

Almost obscured in the flurry of attention Senate Bill 6 received was the fact that, for the first time, both the Legislature and the Board of Governors were aligned in agreement that there should be greater segmentation of Florida's higher education institutions in terms of mission differentiation. In the past, funding formulas and special state appropriations were driven primarily by political decisions, and less on the importance of having a strong higher education system aligned with state goals and workforce needs. This past practice led to having the major research universities viewed as similar in their needs as the small regional universities and state/community colleges. Related to this new emphasis on mission differentiation, the University of Florida has already made a strategic decision that it will de-emphasize (but not eliminate) initial teacher preparation at the undergraduate level and focus more on graduate education, a stance that runs counter to the argument that research universities should become more involved with undergraduate teacher preparation (American Council of Education, 1999; Null, 2009). However, given the increased need for strong research determining the characteristics of highly effective teachers, it may now be more appropriate for research universities to limit their involvement in initial educator preparation and concentrate more of their attention on conducting evidence-based research to inform policy (Grossman, 2008; Zeichner & Conklin, 2005).

With these thoughts in mind, we offer some prognostications on the future, and suggest ways college-based programs can adapt to maintain their professional integrity in preparing the next generation of educators, as well as provide continuing assistance and support to the many teachers already in the field who need more high quality professional development (Borko, 2004; Laursen, 2005; Yendol-Hoppey & Dana, 2010). One saving grace is that a greater concentration on professional development could well lead to a greater emphasis on authentic learning and greater utilization of research, rather than a reliance on an outdated model of staff development still prevalent in many school districts. Webster-Wright

(2009) has argued for a paradigm shift in viewing teachers' continuing education through the lens of "authentic professional learning (PL)" rather than professional development (PD), and noted that:

> In seeking a way forward to supporting professionals in their continuing learning, guidelines are required that are congruent with professionals' authentic experiences of learning yet cognizant of the realities of the workplace with respect to professional responsibilities. Constructive strategies need to be developed to enable change from the current practice of delivering PD to that of supporting authentic PL.
>
> (Webster-Wright, 2009, p. 727)

The greater focus on teachers' continued education is likely to be in tandem with a greater use of online programs to deliver instructional services and coaching. Florida may well presage the future Hess (2009) envisions for teacher education:

> Rethinking the shape of the profession calls for a shift away from the assumption that teacher preparation and training should necessarily be driven by institutions of higher education toward a more variegated model that relies on specialized providers, customized preparation for particular duties, and a "just-in-time" mindset regarding skill development and acquisition.
>
> (2009, p. 456)

Despite Hess's dispiriting vision, it is unlikely that change will come as quickly as he predicts. Even given all the rapid changes Florida's teacher preparation programs have endured over the last decade, many of them are still thriving and enrollments are up. Although the teacher shortage may not be as severe as it was in the past, teachers will always be needed, even as their role in the classroom dramatically shifts. Predicting the future is a risky business, but it is clear that colleges which stay abreast of the latest reforms and retool their curricula and realign resources will succeed in a market-driven world. It is not the future many of us wished for when we began our careers in teacher education, but it is the world we live in now.

Notes

1 This trend is already evident in 15 Southern states, based on a report issued by the Southern Education Foundation (Suitts, 2007).
2 It is possible that members of the Chiles Task Force and the Gallagher Task Force overlapped, but given the fact that they were appointed by governors with separate political affiliations it is not likely. We were unable to locate documents that identified the members of either group.

3 The fact that NCATE is now considering approval of alternative programs is likely to increase pressure on the state universities even more, since they will still be obligated to maintain a traditional teacher education program (with all its mandates) in order to maintain their state support, but will now face even more competition from nationally accredited, alternative programs.

4 This issue is a major discussion topic among the state university education deans. At present, all 11 institutions are still committed to keeping their NCATE accreditation, but continuing budget reductions may impact that commitment in the future.

5 The data finally did hit the media with a story in the *Tampa Tribune* in October 2009, but it attracted little widespread public attention.

6 It was also reported that several Republican legislators had second thoughts as the public outcry against the bill began to build, which was one reason why the two bills were rushed through to preclude any amendments. Had additional committee votes been taken, it's quite probable the bill would not have reached Governor Crist for his veto.

7 This phrase was widely used by advocates and State Legislator aides, but no one would claim attribution for it.

8 Personal communication from university and teacher education lobbyists, but, again, no one would go on record as saying this.

References

American Association of Colleges for Teacher Education. (2010). *The clinical preparation of teachers: A policy brief.* Washington, DC: AACTE.

American Board for Certification of Teacher Excellence. (2010). Retrieved August 12, 2010 from http://www.abcte.org/about-abcte

American Council on Education. (1999). *To touch the future: Transforming the way teachers are taught.* Washington, DC: ACE.

Borko, H. (2004). Professional development and teacher learning: Mapping the terrain. *Educational Researcher 33*(8), 3–15.

Boyd, D., Grossman, P., Hammerness, K., Lankford, R.H., Loeb, S., McDonald, M., Reininger, M., Ronfeldt, M., & Wycoff, J. (2008). Surveying the landscape of teacher education in New York City: Constrained variation and the challenge of innovation. *Educational Evaluation and Policy Analysis 30*(4), 319–343.

Chingos, M. & Peterson, P.E. (2010). Do school districts get what they pay for? Predicting teacher effectiveness by college selectivity, experience, etc. Paper presented at the PEPG conference, June, Harvard Kennedy School, Cambridge, MA.

Darling-Hammond, L. (2006). *Powerful teacher education.* San Francisco: Jossey Bass.

Florida Department of Education. (n.d.). Public schools/districts. Retrieved August 9, 2010 from http://www.fldoe.org/schools/ Florida Senate. (2010). *Bill analysis and fiscal impact statement.* Prepared by the Professional Staff of the Education PreK-12 Committee.

Glazerman, S., Seif., E., & Baxter, G. (2008). *Passport to teaching: Career choices and experiences of American Board certified teachers.* Washington, DC: Mathematica Policy Research.

Gordon, R., Kane, T.J., & Staiger, D.O. (2006). *Identifying effective teachers using performance on the job: The Hamilton Project* (Discussion Paper 2006–01). Washington, DC: The Brookings Institution.

Governor's Commission on Education. (1996–1998). Series/Collection Number S1749, meeting files and tapes, Florida State Library and Archives of Florida Online.

Grossman, P. (2008). Responding to our critics: From crisis to opportunity in research on teacher education. *Journal of Teacher Education 59*(1), 10–23.

Hebda, K. & Pfeiffer, R. (2009). Educator quality update. Paper presented at the spring 2009 meeting of the Florida Association of Colleges of Teacher Education, Orlando, Florida.

Hess, F.M. (2009). Revitalizing teacher education by revisiting our assumptions about teaching. *Journal of Teacher Education 60*(5), 450–457.

Imig, D.G. & Imig, S.R. (2008). From traditional certification to competitive certification: A twenty-five year retrospective. In M. Cochran-Smith, S. Feinman-Nemser, D.J. McIntrye, & K.E. Demer (Eds.), *Handbook of research on teacher education: Enduring questions in changing contexts*, 3rd edition (pp. 886–907). New York: Routledge.

Kumashiro, K.K. (2010). Seeing the bigger picture: Troubling movements to end teacher education. *Journal of Teacher Education 61* (1–2), 56–65.

Ladson-Billings, G. (2006). From the achievement gap to the education debt: Understanding achievement in U.S. schools. *Educational Researcher 35*(7), 3–12.

Lampert, M. (2010). Learning teaching in, from, and for practice: What do we mean? *Journal of Teacher Education 61*(1–2), 21–34.

Laursen, P.F. (2005). The authentic teacher. In D. Beijaard, P. Meijer, G. Morine-Dershimer, & H. Tillmema (Eds.), *Teacher professional development in changing conditions* (pp. 199–212). Dordrecht, Netherlands: Springer.

Levine, A. (2006). *Educating school teachers.* New York: The Education Schools Project.

Levine, H., Belfield, C., Muennig, P., & Rouse, C. (2007). *The costs and benefits of an excellent education for all of America's children.* New York: Columbia University, Teachers College.

National Academy of Science, Committee on Prospering in the Global Economy of the 21st Century. (2007). *Rising above the gathering storm: Energizing and employing America for a brighter economic future.* Washington, DC: National Academies Press.

Noell, G.H. & Burns, J.L. (2006). Value-added assessment of teacher preparation: An illustration of emerging technology. *Journal of Teacher Education 57*(1), 37–50.

Null, W. (2009). Back to the future: How and why to revive the teachers college tradition. *Journal of Teacher Education 60*(5), 443–449.

O'Connell, P., Pepler, D., & Craig, W. (1999). Peer involvement in bullying: Insights and challenges for intervention. *Journal of Adolescence 22*, 437–452.

Peske, H.G. & Haycock, K. (2006). *Teaching inequality: How poor and minority students are shortchanged on teacher quality.* Washington, DC: Education Trust.

Peterson, P. E. (2006). *Reforming education in Florida: A study prepared by the Koret Task Force on K-12 education.* Stanford University, Palo Alto, CA: Hoover Press.

Portes, P. (2005). *Dismantling educational inequality: A cultural historical approach to closing the achievement gap.* New York: Peter Lang.

Smith, S. & Cody, S. (2010). An analysis of annual migration flows in Florida, 1980–2008. Retrieved August 9, 2010 from University of Florida, Bureau of Economic and Business Research website, http://www.bebr.ufl.edu/content/analysis-annual-migration-flows-florida-1980-2008

State Board of Education. (2000). *State Board of Education Rule—6a-5.066—Specific Authority 1004.04, 1004.85, 1012.56 FS. Law Implemented 1004.04, 1004.85, 1012.56 FS. History—New 7-2-98, Amended 8-7-00, 3-19-06*, Florida Administrative Code.

Suitts, S. (2007). *A new majority: Low-income students in the South's public schools.* Atlanta, GA: Southern Education Foundation.

Tuttle, C., Anderson, T., & Glazerman, S. (2009). *ABCTE teaching in Florida and their effect on student performance.* Washington, DC: Mathematica Policy Research.

Webster-Wright, A. (2009). Reframing professional development through understanding authentic professional learning. *Review of Educational Research 79*(2), 702–739.

Yendol-Hoppey, D. & Dana, N. (2010). *Powerful professional development.* Thousand Oaks, CA: Corwin Press.

Zeichner, K.M. & Conklin, H. (2005). Research on teacher education programs. In M. Cochran-Smith & K.M. Zeichner (Eds.), *Studying teacher education: The report of the AERA panel on research and teacher education* (pp. 645–736). Mahwah, NJ: Lawrence Erlbaum Press.

APPENDIX A: SUMMARY OF THE FLORIDA CONTINUING PROGRAM APPROVAL STANDARDS

Standard 1: Core Curriculum

The core curriculum standard relates largely to statutory and regulatory language requiring teacher preparation institutions to produce matrices or curriculum maps showing the content of both subject area and professional education competencies for the Florida Educator Accomplished Practices, reading competencies, ESOL competencies, Subject Area Competencies (for the teacher certification area), as well as additional statutory or regulatory areas, such as higher order mathematics, information on the state system of school improvement and accountability, technology appropriate for the grade, and math computational skills acquisition, among others.

Standard 2: Candidate Performance

The candidate performance standard requires assessment of candidate performance on all of the items in Standard 1: Core Curriculum. In addition, approved programs are required to provide data on admissions, enrollment, and completion, and a disaggregation of those data for those who are admitted through waivers of admission. The state allows a 10% waiver of admission per program, per year, with the caveat that the institution must provide remediation for those who are admitted under the 10% rule. Included in Standard 2, Indicator 3, was new language—"Candidates demonstrate impact on P-12 student learning based on student achievement data within field/clinical experiences."

Standard 3: Continuous Improvement

The continuous improvement standard requires a systematic analysis of unit-wide data to determine where improvements are needed and how programs, faculty, and institutions respond to those needs. Standard 3 requires that an institution review whether the programs they offer are needed or sufficient to fill teacher shortage areas. Standard 3 also requires that institutions review both graduates' and employers' satisfaction with the teacher education program. Graduates are asked about their satisfaction with the program's ability to prepare them for their first year of teaching. The institution also analyzes data supplied by the Florida

Department of Education showing the length of stay of graduates and the percentage of completers who are employed a second, third, or fourth year in one of Florida's school districts. And, finally, the institution is expected to analyze pupil achievement data from completers' first year of teaching to determine if the graduates demonstrate a positive impact on P-12 student learning. Achievement data from the Florida Comprehensive Assessment Test, administered to children in Grades 4 through 10 in reading and mathematics, are the primary measures used and available through special request from Florida's Education Data Warehouse.

APPENDIX B: KEY PROVISIONS OF FLORIDA SENATE BILL 6 (2010)

Performance and Differentiated Pay

- o Provides that greater than 50% of each instructional personnel and school-based administrators' pay would be based on student learning gains and provides that the remainder would be based on the performance appraisal;
- o Provides for differentiated pay based on high priority locations, critical teacher shortage areas, or additional academic responsibilities with continued awards contingent upon student learning gains;
- o Provides for a salary schedule for beginning teachers that is in effect for the first year that the teachers provide instruction in a Florida K-12 classroom;
- o Prohibits districts from using time-served or degrees held in setting pay schedules;
- o Provides that non-instructional personnel would be paid on the basis of performance.

Professional Certification

- o Requires that a teacher from another state demonstrate subject area mastery by the conclusion of the first semester of teaching;
- o Requires temporary certificate holders to pass a subject area test within the first year of the temporary certificate with some extenuating circumstances;
- o Requires the State Board of Education (SBE) to review the subject area exam and reading instruction rigor;
- o Allows the SBE to adopt rules for accepting college credit recommended by the American Council on Education (ACE);
- o Removes the provision for automatic renewal of National Board for Professional Teaching Standards certificate holders, beginning in 2014;
- o Ties renewal to effective or highly effective performance, as determined under the performance appraisal;
- o Prohibits the assignment of a beginning teacher to teach reading, science, or mathematics if the teacher was not certified in reading, science, or mathematics.

State Approved Educator Preparation Programs

○ Eliminates the admissions waiver for up to 10% of students admitted to the programs;

○ Requires continued approval of programs contingent upon learning gains, as measured by state assessments;

○ Revises the requirements for per-service field experiences for student teachers;

○ Requires Educator Preparation Institute (EPI) participants to have evidence of eligibility for a temporary certificate and to complete field experiences, mastery of general knowledge, and subject area testing prior to completion of the program;

○ Requires EPI instructors to meet the same requirements as instructors in traditional teacher preparation programs;

○ Requires the DOE to submit a report to the Governor and the Legislature on the effectiveness of state approved teacher preparation programs;

○ Requires the Office of Program Policy Analysis and Government Accountability (OPPAGA) to review the current standards for the continued approval of teacher preparation programs and make recommendations to the Legislature.

EDITORS' COMMENTARY

Once upon a time education school deans from "flagship" institutions and state superintendents or chief state school officers had breakfast or lunch frequently and aired their hopes and aspirations, gripes and grievances, frustrations and fears. There was cordiality and cooperation, respect and regard. There were common problems to be addressed and both sides made efforts to surmount these problems. This policy case study suggests that a chasm has opened between state elected leaders and those charged with preparing teachers and principals for the schools in Florida. Now education deans are ordered to meetings and chastised for their perceived deficiencies. Bully metaphors are used to describe the actions of state officials and there is talk of gauntlets being thrown down.

The Florida study suggests that higher education based teacher education has served as a convenient scapegoat for politicians and policymakers. When things go wrong with their reform efforts, the authors of this case argue, all too often the blame comes to university-based programs. Somehow, the inequitable distribution of resources to K-12 schools or the growing achievement gap between rich and poor schools is due to the local education school. The authors suggest that this is unfortunate and unfair. In effect, they argue that colleges and schools of education have complied with ever-changing rules and regulations, done so with alacrity and efficiency, and fulfilled their responsibilities to prepare high quality teachers to staff the schools of the state. Florida politicians, they insist, have "responded with 'but that wasn't fast enough nor good enough'."

Seemingly, Florida has always been confronted with a shortage of teachers for its K-12 schools. Task forces and study commissions have been convened and efforts made to get more college graduates into teaching. Governors have brought together experts and academics, teachers and administrators, to address every facet of the problem. Reports are produced and legislative actions taken. At one point, as in Indiana, the response was that to professionalize teaching would result in more candidates for teaching—that higher status and more recognition of teaching as a profession would attract more students to teaching. In response, Florida created a professional standards board and Florida's institutions invested heavily in professional controls—national accreditation, national certification, and a refashioned state licensure process. But this quest for professionalism, the authors assert, soon gave way to the practicality of getting more highly qualified teachers into the classrooms of Florida. Quantity trumped quality as the dominant theme in teacher education.

The case documents the proliferation of alternative routes to licensure in Florida. The authors claim that these alternative routes were created without sufficient evidence as to their potential efficacy and are being held to a set of different expectations than are collegiate-based programs. Although the deans yelled "foul"—and called for the equal treatment of all preparation programs—the events as documented in this chapter do not hold out much hope that policymakers will change their attitudes or invest more in collegiate-based programs. Implicit in this description is an illustration of sense-making co-construction. As the policies from the state were being implemented in community colleges and school districts, the education schools chose to do so as well. Creating high quality alternative programs became the response of some institutions. However, others chose to create a context of being a victim. The real issue that emerges is the disconnect between policymakers and education school leaders, with the deans attempting to find ways to justify the status quo long past a time when the majority of policymakers were willing to accord education schools a special place among the so-called portfolio of providers that state government in Florida came to champion.

The authors plead for a more rational policymaking process. Policy should be based on evidence, they insist, and such evidence should be unbiased and non-political. The authors yearn for evidence that would affirm their presumed preeminent role in preparing the very best teachers for Florida. They recognize the need to build such an assessment system and describe their efforts to now do so by borrowing expertise from their Louisiana colleagues. The conclusion, however, is a disturbing forecast that collegiate-based preparation programs are unlikely to compete with the alternative providers and that education schools must seek new ways to remain relevant in this time of escalating expectations. The answer, for the authors of this case, may be to carve out a niche in meeting the professional development needs of teachers and principals in the state. They acknowledge that in a world of value-added and performance-based assessments of teachers and principals this will be extremely difficult to do.

Discussion Questions

1. Using the tools of sense-making, analyze the response of this education school to the efforts of the state to create alternative routes. Was the appeal for evidence a way to justify the status quo? Given that education schools are situated in universities, what were the constraints imposed by its university setting on the effort to create its own alternative certification program? Can education schools respond to state policies in ways similar to schools and school districts? Should they?

2. The case explores the escalation of expectations for teacher education over the past decade. With the space for and funding of teacher education in decline, how do we justify or accommodate the added expectations for more ELL teachers, more teachers skilled in new forms of classroom management, teachers more able to use new technologies and interpret data, teachers more sophisticated in their approaches to instruction and their commitments to students?

3. The authors argue that if there had been more data, evidence, and information available, policymakers would not have turned to alternative providers to prepare more teachers. What evidence would have been needed to dissuade policymakers from the course of action they took in Florida in the 1990s relative to teacher education?

4. The authors suggest that there are a group of "depressingly familiar" policy influentials who affect policymaking in the States. Can we identify those who have such influence and see the ways that they go about influencing policy?

5. At one point, the assertion is made that some politicians are seeking to close education schools because they are too close to the teacher unions. Do you believe this to be true? Why would this be true? How are education schools connected to the teacher unions? How do such connections enhance or detract from the teacher preparation program?

6. Do you see the parallels between Indiana and the professionalization agenda pursued in Florida? What are the similarities and the differences? What were the sources for the efforts to professionalize teaching and teacher education? Why were these sources unable to realize the goals articulated relative to teacher professionalism?

7. The case highlighted the appeal to professional protections—professional accreditation being the most noteworthy example—yet the story that is told is that those protections were insufficient to overcome the machinations of ambitious governors "with education agendas." What should be expected of professional accreditation? Should those expectations rise above the political demands of state agencies? How are the professional and the state agendas for teacher education program intertwined?

8. Does the need for an adequate supply of good enough teachers always trump the expectations for high quality teachers? Do we know how to distinguish

between good enough teachers and high quality teachers when they are in initial or preservice programs? Given good enough teachers and high quality teachers, which students are likely to get the good enough teachers? How can we address the paradox of claiming high quality when we really mean attracting enough teachers?

9. Can research-extensive universities in high demand states meet the needs and expectations of schools and school districts for beginning teachers or is the mission of these institutions so unique or different from those of other higher education institutions that they should be encouraged to focus on research expectations instead of preparing teachers? What incentives are necessary to help education schools meet these expectations? Should sanctions be imposed if they fail to meet these expectations?

5

ASSESSING STATE AND FEDERAL POLICIES TO EVALUATE THE QUALITY OF TEACHER PREPARATION PROGRAMS

Ken Zeichner[1]

> States like Louisiana are leading the way in building the longitudinal data systems that enable states to track and compare the impact of new teachers from teacher preparation programs on student achievement over a number of years . . . Louisiana is using that information to identify effective and ineffective programs for the first time—and university-based teacher education programs are using the outcomes data to revamp and strengthen their programs . . . Louisiana is the only state in the nation that tracks the effectiveness of its teacher preparation programs. Every state should be doing the same . . . It's a simple but obvious idea—colleges of education and district officials ought to know which teacher preparation programs are effective and which need fixing.
>
> (Duncan, 2009, p. 5)

In this chapter, I examine the warrants for various existing and proposed state and federal policies related to accountability in preservice teacher education programs in the U.S. Given that it is very clear that currently little or no empirical evidence exists that supports the efficacy of particular accountability policies and processes used in state program approval and national accreditation (Wilson, Floden, & Ferrini-Mundy, 2001; Wilson & Youngs, 2005; National Research Council, 2010), I will examine the warrant for specific policies and practices based on a number of other criteria. These include how preservice preparation programs are assessed in other professions, the state of our current methods for assessing teachers' knowledge and teaching skills, the costs and projected benefits associated with particular practices and what are reasonable ways to hold teacher preparation programs accountable for their work. With regard to projected benefits, I will give attention to the likely ability of an accountability practice both to illuminate the quality of teacher education programs and contribute to improving programs.

Although the analysis will discuss a number of different policies and practices,[2] I will give particular attention to two accountability practices that are under intense discussion in the current policy context: the development of a rigorous teacher performance assessment that would be used for completion of a pre-service program and initial teacher licensing and, as noted in the quote above, the evaluation of the quality of a teacher education program based on a value-added analysis of the standardized test scores of elementary and secondary school pupils taught by graduates of specific programs. The latter practice, referred to as the "positive impact mandate" (Hamel & Merz, 2005, p. 158), has received extensive and largely uncritical coverage in the national print and broadcast media (e.g., Abramson, 2010; Glenn, 2010). Both of these practices have been endorsed by the current federal education department (e.g., Duncan, 2010) and it is important that they receive a careful examination while their implementation is still limited. Since the Secretary's talk in October 2009, several other states (e.g., FL, TN. TX, DE) have made moves to implement the positive impact mandate as a form of teacher education program accountability. Federal funding streams such as Race to the Top encouraged more states to join the effort. Given that a new and widely disseminated report was released on accountability in teacher education while this chapter was being written (Crowe, 2010), I will specifically comment on the recommendations made in that report.

Government Policies Related to the Quality of Teacher Preparation Programs

In the last 30 years, both state education departments and the federal government have enacted various policies aimed at assessing the quality of teacher preparation programs that prepare teachers for initial certification. Until the reauthorization of Title II of the Higher Education Act in 1998, and the Elementary and Secondary Education Act in 2001, and despite efforts by the federal government to encourage particular forms of teacher education by providing competitive funds for the use of certain practices (e.g., Clarke, 1969; Earley, 2000a), it was mostly the states and not the federal government that formulated policies and regulations regarding accountability in teacher education (Bales 2006; Imig & Imig, 2008).

Prior to the 1980s, states emphasized an input driven model of program evaluation and approval that judged the degree to which teacher education programs contained the components that were required in a particular state either in terms of opportunities for teacher candidates to study particular topics, the presence of required courses (e.g., a course in teaching reading), or the required number of credit hours devoted to particular topics (e.g., nine credit hours in literacy teaching). These requirements also typically included a minimum number of hours that had to be spent in clinical experiences prior to a full-time student teaching or internship experience and a required minimum number of hours for the full-time teaching experience. For many years, states have licensed individual teachers based on their completion of a state-approved teacher education program (Cronin, 1983; Darling-Hammond et al., 2005).[3]

During my first encounters with the state program approval process in Wisconsin in the 1970s, the process, which occurred every five years, consisted of state education department staff and several K-12 educators auditing syllabi of required courses in teacher education programs to see that the required topics were listed, and checking that the required number of minimum credit hours or time periods for different program components like student teaching or academic minors were in existence. During this period, most states left it up to teacher education institutions to make judgments about the quality of candidates' teaching. The effectiveness of candidates in the classroom was usually judged solely by the observation-based assessments made by college and university-based and school-based supervisors and school-based mentors and administrators, a practice which has been shown to be highly unreliable for measuring teacher effectiveness (Porter, Youngs, & Odden, 2001; Chung Wei & Pecheone, 2010; Wilson, 2009).

Unlike some other countries where there are national standards related to licensure and program quality (Wang et al., 2003), individual states in the U.S. set their own policies. There is some degree of overlap in state requirements however, as a result of voluntary national accreditation requirements (www.ncate.org or www.teac.org),[4] and consortia of states that have agreed on the use of a number of common standards with regard to teaching and teacher education programs (www.nasdtec.org, www.csso.org). Despite these areas of overlap, individual states' ability to set their own policies with regard to accountability for teacher education programs has resulted in accountability and licensing requirements that have been called "haphazard" in the most recent report on teacher quality by the U.S. Secretary of Education (U.S. Department of Education, 2009).

Beginning in the 1970s in several southern states, and then moving to other areas of the country in the 1980s, states began to require a variety of tests to enter and complete teacher preparation programs and generally became more prescriptive about the teacher education curriculum (Cronin, 1983; Prestine, 1989).[5] These tests include basic skills tests (currently 27 states),[6] tests of professional knowledge and pedagogy (currently 28 states), and tests of core academic subject-matter content (currently 37 states) and other subject-matter content (currently 32 states) (NASDTEC, 2010). Currently there are about 1,100 different tests used for initial teacher licensure throughout the U.S., with each state choosing its own tests and setting its own passing scores (Crowe, 2010). According to a report by the Education Trust (Brennan, 1999), most state subject-matter licensure tests are viewed as too easy and not relevant to ensuring that teachers have the academic skills that they need to be successful in raising student achievement. They also have very little predictive validity with regard to future success in teaching (Goldhaber, 2010; Wilson & Youngs, 2005). Despite these and other criticisms about the value of the current teacher tests (Berry, 2010), pass rates on teacher licensure tests are used as a component of the accountability system in 32 states.

For example, in New York, 80% of program completers from individual programs must pass the required tests for the programs to avoid sanctions by the state (NASDTEC. 2010).

During the 1970s, states began to introduce performance assessment in teacher education, and competency-based or performance-based teacher education (C/PBTE) was required or there were plans to require it in over 20 states for program approval and the initial licensing of teachers (Gage & Winne, 1975). At one point, all National Teacher Corps projects were required to use performance-based assessment (Houston & Howsam, 1972) and the American Association of Colleges for Teacher Education encouraged teacher education programs to become competency-based. AACTE provided a wealth of resources to help them do so in specific aspects of their programs such as multicultural education (e.g., AACTE, 1974). C/PBTE[7] was advocated as an alternative to making teacher education programs accountable according to whether they contained all the state required coursework and fieldwork. Uncoupling courses and credits from state licensure requirements with C/PBTE was supposed to enable programs to innovate and to develop different approaches, and was a key factor in the movement toward alternative routes to teaching (Sykes & Dibner, 2009).

For a variety of reasons, including the cost of implementation and the lack of solid research supporting the connections between teacher competencies and student learning (e.g., Heath & Nielson, 1974), C/PBTE temporarily disappeared from U.S. teacher education with the exception of a few states like Florida and Georgia and programs like Alverno College in Milwaukee (Zeichner, 2005). Around 2000, C/PBTE once again gained momentum in teacher education accountability communities in the U.S. with the adoption of performance-based assessment by NCATE and the implementation by some states of performance standards for initial teacher licensing and program approval (Valli & Rennert-Ariev, 2002). State teaching standards in 16 states in this current incarnation of C/PBTE are based in part on the standards developed by the Interstate New Teacher Assessment and Support Consortium (INTASC) that is a part of the Council of Chief State School Officers (www.ccsso.org).

For example, while I was working in 2004 in Wisconsin, state program approval shifted from a system that focused only on program inputs (e.g., Are the required topics and credits in the teacher education curriculum?) to an accountability system that emphasizes performance-based assessment of teacher candidates. Some states like Wisconsin examine the quality of the performance assessment systems in teacher education institutions for program approval, while other states like Washington also want to see evidence in candidate performance assessments that teacher candidates have achieved a certain level of competency on the state teaching standards. Currently, approximately 19 states require a performance assessment of teaching for initial licensure (NASDTEC, 2010).[8]

There have been various responses by teacher educators to the shift toward performance-based assessment as a part of initial teacher licensure and state program

approval. On the one hand, there is a concern that performance assessment nega-
tively affects the ability of teacher educators to engage in the practices that they
think are needed to educate beginning teachers well by diverting the attention of
teacher educators and the limited resources of their institutions to activities they
perceive as not related to their core mission (Berlak, 2010; Kornfeld et al., 2007;
Rennert-Ariev, 2008). On the other hand is the argument that teacher performance assessment data (unlike value-added assessment data) potentially provide
teacher educators with useful information that they can use in improving their
programs (Peck, Gallucci, & Sloan, 2010) and serves as a form of learning for
teacher candidates (Darling-Hammond, in press; Diez & Haas, 1997; Chung Wei &
Pecheone, 2010). Mostly, however, C/PBTE has not been fully implemented in
many teacher education institutions despite state requirements, because of the costs
and other issues associated with a genuine implementation of the idea (Zeichner,
2005). State departments of education have experienced cuts in their budgets and
staff over the years and do not have the capacity in many cases to monitor and
enforce a genuine performance-based system (Darling-Hammond, 2005).

What is a Reasonable Approach to Accountability in U.S. Teacher Education?

Teacher Licensure Exams and System Coherence

Given the lack of empirical evidence related to particular accountability policies
and processes in teacher education, one way to begin to formulate a position on
an accountability system for teacher education is to look at how the quality of
other professional schools are assessed. When one examines how other professions
evaluate the readiness of individual candidates to practice and assess the quality of
the preservice programs that prepare them, it is clear that there is much more
uniformity across the country with regard to how other professionals are licensed.
Crowe (2010) and Neville, Sherman, and Cohen (2005) discuss licensing and
program approval requirements in a number of professions such as medicine, law,
accountancy, nursing, and engineering, and all of these other professional schools
have either a national licensing exam or a state exam with a national component
before candidates are allowed to practice. Some professional schools also use
performance assessments, and the structured observation and evaluation of clinical
practice.

Crowe (2010) calls for both a major overhaul of teacher licensing exams and
greater uniformity across the nation in teacher standards, policies, and program
approval processes. Both of these recommendations are reasonable ones given the
practices in other professions. As Berry (2010) points out, however, merely raising
the cut scores on current teacher licensing exams, as some have suggested, will not
necessarily lead to improvements. For example, research by Goldhaber (2007)
showed that raising the cut scores on the North Carolina licensing exam up to the
level used in Connecticut would eliminate teachers who have proven that they

can produce higher student achievement on standardized tests. Other analyses have shown the disproportionate failure rates on some exams by minority teacher candidates (Gitomer, Latham, & Ziomek, 1999; Villegas & Davis, 2008). Crowe's (2010) recommendation to engage in a major overhaul of teacher licensing exams and to make them more uniform in content and cut scores across the nation seems warranted as a general recommendation.

We have to keep in mind, though, the growing empirical evidence related to the importance of building a more ethnically and racially diverse teaching force in terms of its positive impact on student learning, particularly learning for students of color (Villegas & Davis, 2008). We also have to remember that the purpose of initial licensure tests is to separate those candidates who are minimally competent from those who are not. The National Research Council (NRC) report on teacher testing concluded that "a set of well designed tests cannot measure all of the prerequisites of competent beginning teaching" (Mitchell et al., 2001, p. 165). This group concludes that multiple measures of beginning teacher effectiveness are needed and that decisions about licensing should not be made on licensure tests alone. So, while Crowe's (2010) recommendation that we need to apply higher standards in a new set of teacher licensure tests that are more uniform across the nation makes sense up to a point, there are real dangers in raising the cut scores too high. "Setting substantially higher passing scores on licensure tests is likely to reduce the diversity of the teaching applicant pool" (Mitchell et al., 2001, p. 167) and, as Goldhaber's (2007) research noted above concluded, keep potentially effective teachers out of the classroom.

Assessments of Teacher Effectiveness in the Classroom

Throughout the history of formal American teacher education programs, teacher candidates have had to demonstrate their competence in a classroom as part of program completion (Fraser, 2007). Throughout much of this history, these judgments were made by school-based or college and university-based supervisors and mentors based on brief classroom observations. The unreliability of these assessment measures of teaching quality has been demonstrated in the literature (e.g., Chung Wei & Pecheone, 2010; Porter, Youngs, & Odden, 2001; Wilson, 2009). Crowe's (2010) recommendation that accountability systems in teacher education "should include a measure of teacher effectiveness that reports the extent to which program graduates help their K-12 students to learn" (p. 12) is a reasonable one that can be approached in a number of different ways. One way to obtain assessments of teachers' ability to promote student learning is to strengthen the weak systems of student teacher assessment that exist in many clinical preparation experiences across the nation.[9]

When I began my career as a university teacher educator in the 1970s, efforts were made to infuse some of the more structured classroom observation instruments into student teacher/intern supervision (e.g., Simon & Boyer, 1974) and to build a body of research and sound practices in supervising clinical experiences in teacher education (e.g., Goldhammer, 1969). The goal in these efforts was to raise the quality of the mentoring and assessment of teacher candidates during their clinical experiences with a focus on students and their learning.

Today there is wide consensus that the quality of supervision and assessment in clinical experiences in preservice teacher education is highly uneven (AACTE, 2010). Currently, there are a number of efforts like the Working with Teachers to Develop Fair and Reliable Measures of Effective Teaching project funded by the Bill and Melinda Gates Foundation (http://www.gatesfoundation.org/highschools/Documents/met-framing-paper.pdf) to develop higher quality classroom observation-based assessments of the quality of teaching. Other notable efforts to make direct assessment of actual teaching in the classroom a central feature of educational accountability include the Classroom Assessment Scoring System or CLASS (Pianta & Hamre, 2009) and an observational framework based on the ETS Praxis III performance assessment (Danielson, 1996). Improving the quality and consistency of supervisor and mentor teacher assessments of teacher candidates is an important part of a strategy to measure the effectiveness of teachers in the classroom before giving them an initial teaching license or allowing them to serve as teachers of record. Using observational frameworks designed for research purposes in classroom-based assessments in preservice clinical experiences will require some adaptations, but we know from adaptations of parts of systematic observation instruments during the era of teacher effectiveness research in the 1970s that this is a doable task (e.g., Acheson & Gall, 1980).

It is not very common for either college- and university-based field supervisors or school-based mentors to be required to receive preparation for their work as supervisors of teacher candidates. In fact, the P-12 teachers who provide the bulk of mentoring and assessment of teacher candidates in most programs rarely receive the compensation and support that are justified by the important role that they play and the time that they spend on this work in many teacher education programs (Zeichner, 2006).[10] Improving the consistency and the quality of field supervision for teacher candidates should be a priority in efforts to raise the quality of how we assess the quality of teacher candidates' teaching.

Another strategy for including a measure of teaching effectiveness that includes the ability to be successful in achieving student learning as a part of initial licensing is to utilize a high quality teacher performance assessment. Berry (2010) and Darling-Hammond (2009, in press) lay out a convincing case for the use of such an assessment based on research evidence from the beginning teacher assessments in Connecticut, and the National Board assessments (also see Darling-Hammond & Chung Wei, 2009). During the last several years, researchers at Stanford have led the development of a rigorous teacher performance assessment (Performance Assessment for California Teachers or PACT) that is used in over 30 California teacher education institutions. Despite some concerns about the assessment and about the lack of funding to support its implementation (Berlak, 2010), this assessment has been shown in some cases to be able to predict teacher effectiveness according to student learning and to support teacher learning and teacher education program improvement (e.g., Chung Wei & Pecheone, 2010; Newton, Walker, & Darling-Hammond, 2010; Pecheone & Chung, 2006; Peck, Gallucci, & Sloan, 2010).

AACTE and the CCSSO are currently supporting a project involving 20 states that is developing a nationally available performance assessment based on the PACT that meets high standards of reliability and validity and that can be used in a variety of states for candidates to demonstrate their mastery of state teaching standards.[11] This assessment combines embedded signature assessments in individual teacher education programs with a capstone teaching event used across all institutions (Diez, 2010). The capstone teaching event which is usually done in the final student teaching or internship experience engages candidates in documenting their practice in relation to academic language, planning, teaching, assessing, and reflecting according to a set of guiding questions and structures. The responses of the candidates are then evaluated by trained scorers according to a set of carefully designed and field-tested rubrics. Extensive reliability and validity studies have been, and continue to be, carried out on the PACT (Chung Wei & Pecheone, 2010; Pecheone & Chung, 2006) and on the new nationally available performance assessment that is based on it.[12]

Recently, I had my first direct experience with the PACT in the elementary and secondary teacher education programs at the University of Washington-Seattle that I currently direct. The implementation of this assessment is understandably a more complicated and expensive enterprise than what currently exists in most programs, with the need for scorer training, building opportunities to learn into the teacher education curriculum, coordinating the assessment with the placement schools and so on (a good assessment requires resources), and the kind of data about our candidates' teaching that emerged from this assessment on the performance of our teacher candidates was invaluable. For example, in our secondary program, we devoted several program meetings to discussions of various forms of the teacher performance assessment data (including artifacts from the assessment) that involved both university- and school-based teacher educators. These discussions led to revisions in the program curriculum for the next cohort of candidates. For example, a number of our secondary teacher candidates scored low on the academic language component of the assessment and some revisions were made in the curriculum and candidate assignments to address these areas of weakness.

There are other performance-based assessments of teaching besides PACT, such as the ETS Praxis III assessment[13] and the protocols for evaluating candidate work samples developed originally at Western Oregon University (McConnery, Schalock, & Schalock, 1998).[14] The goal of the CCSSO and AACTE project is to develop a more uniform approach to performance-based assessment in teacher education than the current approach of allowing each state to choose what assessments they will use. Even if it turns out that there is more than one performance assessment used by states, there should be a requirement that all of the assessments used to assess the quality of teaching effectiveness of program completers meet a set of common standards with regard to their psychometric quality.

The use of portfolios for the assessment of the quality of teacher candidates' teaching is widespread in U.S. teacher education programs (Delandshere & Petrosky, 2010), but most of the portfolios that are used are relatively unstructured compared to the PACT and the Teacher Work Sample Methodology developed at Oregon State and do not have the psychometric quality to be used effectively as a summative assessment tool (e.g., Chung Wei & Pecheone, 2010; Wilkerson & Lang, 2003).

Finally, another way to assess the teaching effectiveness of teacher candidates after they complete their preparation programs, and to supposedly judge the quality of these programs, is to use value-added analysis (VAA) to link growth in standardized test scores of pupils to the programs from which teachers graduated and then to rank teacher education programs in each state according to the alleged contribution of their graduates to student learning. Currently, as pointed out earlier, the national media have been obsessed with this strategy (e.g., Abramson, 2010; Glenn, 2010; Honowar, 2007; Kelderman, 2010) and the Secretary of Education, as illustrated in the opening quote of this chapter, travels the country promoting the idea. Louisiana is continually identified as the model for other states to follow in this area (Noell & Burns, 2006) along with Florida that has already begun ranking teacher education programs according to the value-added test scores of pupils taught by graduates from the different programs in the state (Glenn, 2010).

In the last few years, there has been much debate about the wisdom of using VAA to tie growth in students' standardized test scores to specific teachers and teacher education programs. For example, researchers have shown that using value-added student achievement scores to measure teaching effectiveness requires at least three years of data (McCaffrey, Lockwood, Mariano, & Setodji, 2005). Other researchers have questioned the assumptions on which VAA models are based and warn about their careful use (Rothstein, 2010). Some researchers have also raised questions about the tests that are used as measures of student learning (Darling-Hammond & Chung Wei, 2009). Finally, because the results one gets in VAA vary according to the decisions researchers make about how to handle the data, there is wide consensus that VAA should not serve as the sole basis for making decisions about teachers (Braun, 2005).

The National Research Council (2010) report on teacher education in the U.S. examined the relevance of VAA for evaluating teaching and teacher education programs. The report acknowledges some of the concerns that have been raised about this method, including:

> That value-added methods do not adequately disentangle the role of individual teachers or their characteristics from other factors that influence student achievement . . . there are concerns about measures of student outcomes and accurate measurement of teacher education attri-

butes . . . Another concern is that student achievement tests developed in the context of high stakes accountability goals may provide a distorted understanding of the factors that influence student achievement.

(2010, p. 29)

After acknowledging these and other concerns, the report concludes:

As with any research design, value-added models may provide convincing evidence or limited insights depending on how well the model fits the research question, and how well it is implemented. Value-added models may provide valuable information about effective teacher preparation, but not definitive conclusions and are best considered together with other evidence from a variety of other perspectives.

(2010, p. 29)

Very few of those who have advocated the use of VAA to evaluate the quality of teacher education programs have advocated their use alone as a measure of effectiveness, including Crowe (2010). Levin (1980) advocated the use of a cost-utility analysis for evaluating the wisdom of using particular components in both teacher licensing and teacher education program accountability systems. When one follows Levin's advice, the question arises as to whether it is worth the time and expense to gather value- added data for the purposes of program accountability given the lack of consensus about the wisdom and/or the reasonableness of doing so, and the questionable quality of the information it provides. Couldn't a rigorous and consistent system of teacher education accountability be created that pays attention to teachers' abilities to teach students effectively, utilizing all of the other ways to assess teacher education program quality discussed above and implementing them in a more consistent and rigorous manner than is currently the case?

There are several arguments that should be raised and at least discussed related to the wisdom of VAA as a component of teacher education accountability. None of the popular press articles in the *Chronicle of Higher Education,* or *Education Week* (e.g., Glenn, 2010; Honowar, 2007), the piece on National Public Radio (Abramson, 2010), or articles in local newspapers (Matus, 2009), discusses the concerns that scholars have raised about the methodology in any detail or the fact that scholars disagree about whether and/or how it should be used. They also do not discuss the reasonableness of the approach as a way to evaluate professional schools. In one of the recent statements about the recent Center for American Progress report, it is implicitly asserted that, unless a state is using VAA to evaluate and rank its teacher education programs, it does not "actively hold teacher preparation programs accountable for the effectiveness of the teachers they produce" (Center for American Progress, 2010, p. 1).

The first question that should be asked about VAA is whether there are any other professional schools evaluated in this way on the basis of student/client/patient outcomes after the candidates have completed their preparation programs. Using VAA as a required component of a teacher education accountability system would be analogous to evaluating medical schools according to how well graduates of particular medical schools were able to help particular patients get well, or how many cases graduates of particular law schools won or lost, or how many clients of accountants from particular business school programs were audited by the IRS, and so on. While Crowe (2010) and other critics of teacher education are eager to draw on the accountability systems for other professional schools to advocate for more uniformity in practice, no one has mentioned the fact that there is not a single profession where preparation programs are held accountable in the accreditation process for student/client/patient outcomes beyond the point of graduation. As has been pointed out in discussions of accountability in other professions (e.g., Neville, Sherman, & Cohen, 2005), uniform licensing exams that sometimes include a performance assessment component are standard practice for assessing candidates' readiness to practice and for assessing the quality of medical preparation. To require that teacher education programs be held to a standard of accountability that no other professional school is held to require is a practice for which a justification has not been provided.

A second question that should be raised about the wisdom of using VAA for teacher education program accountability is the usefulness of the data that it provides about the elements of teacher education programs that are related to positive results. Despite the fact that teacher education program administrators are often quoted in popular press articles promoting the use of VAA results in stimulating program improvement (e.g., Matus, 2009), the fact is that a ranking of institutions using VAA provides very little, if any, information about the particular features of programs that are linked to the outcomes. Although there are a few examples of research projects that use VAA in combination with other methods to illuminate the particular features of teacher education programs that are linked to positive and negative outcomes (Boyd et al., 2008), the kind of analyses that have been produced to date in Louisiana and Florida are not sophisticated enough to produce data that illuminates the particular aspects of preparation programs or teachers' practices that would be useful to program improvement. On the other hand, as has been discussed earlier, there are examples of how specific information about candidates' teaching from a rigorous teacher performance assessment can be used to support program renewal and improvement (e.g., Peck, Gallucci, & Sloan, 2010).

The comparative costs of implementing a VAA-driven accountability system and of developing high quality teacher performance assessments to be used for program completion, together with the value of the data produced for stimulating program renewal and improvement, suggests that strengthening classroom observation-based assessment and developing high quality performance assessments are much more worthy activities to undertake than investing in VAA to assess the teaching effectiveness of teacher education program graduates.

Although Harris and McCaffrey (2010) argue that, given the current system of standardized testing in the U.S., the cost of creating value-added (VA) measures is quite low, they also acknowledge that the costs associated with calculating the VA measures are only a part of what is needed to adopt a VAA approach to evaluate teaching. In addition to calculating the actual measures, they argue that educators need to be trained in how to use VA measures and to understand their limitations and investments need to be made in overcoming some of the technical problems that have limited the usefulness of VA data to date.

Discussion

In this chapter, I have briefly discussed a number of existing and proposed policies and processes for strengthening the system of teacher education accountability in the U.S. While I have supported certain specific recommendations and general principles advocated in the Center for American Progress's recently released report on teacher education accountability (Crowe, 2010)—such as engaging in a major overhaul of teacher testing, creating greater uniformity throughout the nation in policies and practices, using high quality assessments of candidates' teaching as part of initial licensure and program approval, and holding all teacher education programs to the same accountability standards—I have argued against the ranking of teacher education programs based solely on the VAA of pupil test scores of their graduates as a reasonable, cost-effective and useful way to assess candidates' teaching effectiveness and to evaluate the quality of teacher preparation programs.

No other professional school is held accountable for student/patient/client outcomes after program completion in this way, and the data that are produced by VAA, lacking information about the specifics of teaching and the contexts in which it takes place, do not contribute to the improvement of teaching or teacher education programs. It would be a wiser strategy to invest in improving both classroom observation-based assessments in clinical experiences and to develop a high quality teacher performance assessment to be administered at the completion of a pre-service program. Both of these types of assessments would provide much more specific information about the ability of teacher candidates to effectively produce student learning and, although they are expensive, they will have a much greater impact on improving the quality of teacher education in the U.S. than a VAA approach.

How to Achieve Greater National Uniformity

There are different ways in which we can move toward greater uniformity in initial teacher licensure and program accountability throughout the U.S. Some (e.g., Darling-Hammond & Baratz-Snowden, 2005) have argued for mandatory national accreditation of teacher education programs pointing to mandatory national accreditation of preparation programs in other professions. It is also potentially possible to bring about greater consistency in state requirements and policies by voluntary cooperation among state education departments and professional standards boards. Given the criticisms that have been leveled at the bureaucratic nature of national accreditation of teacher education in the past (e.g., Johnson et al., 2005)[15] and the fact that there is little existing empirical evidence as to the value of national program accreditation (National Research Council, 2010; Wilson & Youngs, 2005), the voluntary cooperation approach advocated by Crowe (2010) seems like a reasonable approach at least for now.

One can question, though, the likelihood of states voluntarily agreeing to adopt the same licensure standards. In the long run, some form of mandatory national program accreditation will probably be the only way to achieve a more uniform national accountability system for teacher education in the U.S. Priority should be given to the National Research Council's (2010) recent recommendation to undertake an independent evaluation of program accreditation in teacher education. This evaluation could lead to a revision of the current system and then to a requirement that all teacher education programs be nationally accredited. If the redesigned system is streamlined and made more manageable and cost effective, is meaningful in the sense of getting at actual program quality,[16] and helps contribute to the improvement of programs, it will likely be positively received by teacher education institutions.

It is very interesting how commentators like Crowe (2010) draw on other professions in a very selective way. Although they draw on accountability in other professions as a reason for creating greater uniformity in teacher education accountability, they fail to point out that in most of the other professions that are used as illustrations the profession itself plays a significant role in setting and enforcing accountability standards. Whatever accountability system is developed for teacher education in the U.S. must include a significant role for the profession in setting and enforcing standards for teachers and teacher education programs along with greater national uniformity.

High Quality Teacher Education Accountability is Expensive

Another issue that needs to be faced in the creation of higher quality standards for initial teacher licensure and program accountability is that a higher quality system will cost more than what is currently in place. The National Research Council report (2010) on teacher education in the U.S. and the most recent teacher quality report by the U.S. Secretary of Education (U.S. Department of Education, 2009) indicate that somewhere between 70% and 85% of new teachers today have been prepared by a college and university program of some kind. Given the consistent decline over a number of years in state support to the public universities where most college and university educated teachers are prepared (Lyall & Sell, 2006), as well as the continuing cuts in the budgets of the state education agencies that would implement a substantial part of a strengthened accountability system alone or in conjunction with national accreditation bodies (Darling-Hammond et al., 2005), the question of how a higher quality accountability system in U.S. teacher education will be funded is a serious issue that needs to be resolved.

One strategy used in the past to fund components of the accountability system in teacher education has been to shift the costs to prospective teachers. However, with the widespread use of teacher testing throughout the country at various points in initial licensing and the sharp rise in college and university tuition to offset reductions in state funding, it has become quite expensive for teacher candidates to meet state requirements for initial licensure.[17] The implementation of a high quality teacher performance assessment with good reliability and validity and strengthening classroom observation-based assessment during clinical experiences will also be very expensive. This past year, for example, to administer a version of the PACT to about 130 elementary and secondary teacher candidates at the University of Washington-Seattle, we spent approximately $35,500 for the training of scorers, and paying scorers for initial scoring and rescoring (about $273 per candidate). These costs do not include the salary of a half-time staff person to coordinate the whole process and the substantial staff and faculty time that were spent in designing the infrastructure that was needed to support the assessment and to integrate it into the teacher education curriculum. The state of Washington now has a requirement for an evidence-based performance assessment in all its teacher education programs and discussions are taking place about how the costs of implementing this assessment will be paid. Shifting the costs to teacher candidates who are already paying higher tuition and fees for required basic skills and content exams is problematic given the negative effects this is likely to have on the goal of building a teaching force in the state that is more representative of the population in the state.

When one applies Levin's (1980) cost-benefit analysis approach to the problem of assessing the teaching effectiveness of teacher candidates as a component of teacher education program accountability, the most expensive option in the range of alternatives is the implementation of VAA to rank teacher education programs in each state. The money that would be spent in implementing these analyses in every state, and in training people to use them, could more wisely be spent on supporting directions for reform in teacher education that research shows makes a difference in producing high quality teacher education programs, such as: strengthening the clinical component of preparation and its connection to the rest of the program (e.g., Boyd et al., 2008), supporting the development of higher quality classroom observation and performance assessments, and supporting research and evaluation on teacher education including the needed evaluation of the accountability system in teacher education. The return on investment on these and other expenditures would be greater than that from investing in an expensive VAA accountability system that provides very little data about teaching that can be useful for improving teaching and teacher preparation programs.

Identify and Punish the "Culprits" vs Help Programs Become Better

Sykes and Dibner (2009), in their review of federal policy related to teaching in the U.S. over the last 50 years, make a distinction between sanctions-oriented policies that are designed to identify and punish the culprits (Earley, 2000a) and accountability that is designed to contribute to the improvement of teachers, schools, and teacher education programs. They argue for policies that provide useful data that contribute to the improvement of teaching and teacher education. There is a certain cynicism among a number of critics of the current teacher education accountability system about the intentions and motives of teacher educators in colleges and universities and there are even accusations in some cases that teacher educators are trying to get away with something dishonestly. A statement by Crowe (2010) is a good example of this cynical attitude. Referring to the 1998 reauthorization of Title II of the Higher Education Act and the required state report cards on candidate pass rates on content exams. Crowe states:

> Shortly after the report card statute was established, a significant number of institutions and state agencies joined with the teacher education professional associations—the American Association of Colleges for Teacher Education or AACTE as well as NCATE—to work out a way to beat the reporting system. The trick they devised was requiring teacher candidates to pass all required teacher tests before being allowed to graduate. This allowed programs to report 100 percent pass rates on the teacher tests.
>
> (2010, p. 9)

This statement about "trickery" which criticizes how teacher education institutions responded to requirements reporting on tests that Crowe (2010) has concluded is essentially bankrupt is very interesting.[18] This attitude of "somebody is trying to get away with something" raises a question about what the purpose of the tests are in the first place. Isn't the goal of requiring candidates to pass licensing exams to ensure that those who receive initial teaching licenses have mastery of basic skills and subject areas in their certification areas at a certain level of competence? Hasn't this goal been met if teacher preparation institutions do not recommend candidates for initial licensure if they fail to pass the tests?

It seems to me that developing a fair and rigorous system for monitoring the quality of teacher preparation that closes the weakest traditional and alternative programs and that contributes to the improvement of most programs may not be the real goal of some education school critics who advocate for the use of VAA in teacher education. As Diez (2010) argues, too much emphasis on *proving* that teacher education programs work or don't work can stand in the way of *improving* *them*. What we should be seeking in an accountability system in teacher education is to get an in-depth and accurate reading of the quality of the teachers that programs are recommending to the state for initial certification and a system that contributes to ongoing improvement of preparation programs.

The idea of publicly ranking teacher education in a state according to teacher candidates' performance on licensing exams is not a new idea, as can be seen from Table 5.1, which shows the ranking of teacher preparation institutions in Wisconsin in 1863 in the *Wisconsin Journal of Education*. Although it is a reasonable expectation to hold teacher education programs accountable for the performance of their graduates

TABLE 5.1 Summary of Averages

	Whole No. Examined	Int Arithmetic	Written Arithmetic	Algebra	El Sounds	Spelling	Analysis	Grammar	Composition	Reading	Geography	Physical Geography	Physiology	History	Theory and Practice	Penmanship	Total
Racine High School	6	78.3	87.5	81.7	78.3	86.7	61.7	69.2	60	85	66.7	73.3	75.8	72.5	66.7	75	74.4
Lawrence University	7	75.5	86.4	73.5	68.5	86.4	75.5	77.1	71.4	78.5	56.4	62.9	73.5	65	72.1	77.1	73.3
Allen Grove Academy	13	77.7	76.2	75.5	78.8	80.4	66.5	67.7	52.3	83.1	73.5	68.1	66.2	62.3	62.3	79.6	71.3
Platteville Academy	11	76.8	78.2	64.5	76.3	86.3	60	70.9	62.7	83.2	55.9	63.2	55.4	48.2	65.9	79.5	68.5
Wisconsin Female College	6	72	78.3	67.5	48.3	69.2	73.3	65.8	65	78.3	63.3	67.5	65	63.3	75	62.5	67.6
Fond du Lac High School	4	72.5	72.5	83.7	71.2	81.2	51.2	68.7	63.7	76.2	57.5	50	65	41.2	68.7	80	66.9
Evansville Seminary	20	68.5	74	61	72.7	74	53.2	56.7	57	72.5	61	64.2	62	66.2	70.7	75.5	65.9
Milton Academy	15	76.3	77	58.7	63	76	43.7	59.3	69.3	82.3	52.3	55.3	55	58.3	65	78.7	64.7
Oshkosh High School	3	70	73.3	68.3	70	78.3	46.7	68.3	70	76.7	55	41.6	61.6	50	63.3	71.7	64.3

on licensing exams at the time of program completion, this kind of general ranking of institutions tells us very little about the quality of teaching of the graduates from these institutions. At least it provides some useful information, though, about the relative performance of candidates in the various subject areas covered in the exams that can be used as the basis for examining particular areas of the curriculum.

Table 5.2 shows a recent ranking of education schools in the state of Florida, published in the *St. Petersburg Times* in November 2009. The article begins with a sentence in large type that states that a large local university, the University of Southern Florida, "comes in ninth of the 10 schools when the Florida Comprehensive Assessment Test is used to measure graduates."

This ranking attempts to report on the quality of the ten Florida education schools based on the math and reading scores of students taught by the graduates of different programs. It determined what percentage of graduates from each program had 50% or more of their students make a year's worth of progress. In addition to the fact that this use of value-added analysis did not meet even minimum standards for the use of the method, such as using at least three years of test data, using VAA in combination with other measures of effectiveness, etc. (Berry, 2010; National Research Council, 2010), there is very little useful information provided in this ranking that can be used for program improvement. One could argue that the crude ranking of teacher preparation institutions in 1863 is a more reasonable and useful form of accountability that what was done in Florida in 2009.

The NRC examination warning about the dangers of oversimplification in publically ranking teacher education institutions based on licensure test scores can also be applied to the ranking of institutions by VAA scores alone.

The public reporting and accountability provisions of Title II may encourage erroneous conclusions about the quality of teacher preparation.

TABLE 5.2 Rating Teacher Preparation Programs

University	Percentage of Teachers with 50% or More of Students Making Learning Gains	% "High Performing"*
Florida A&M	80	7
Florida Atlantic	84	19
Florida Gulf Coast	77	14
Florida International	85	23
Florida State	81	20
University of Central Florida	83	20
University of Florida	84	18
University of North Florida	84	11
University of South Florida	76	15
University of West Florida	70	11

Note: * Based on FCAT learning gains that were particularly large.

Although the percentage of graduates who pass initial licensure tests provides an entry point for evaluating an institution's quality, simple comparisons among institutions based on their pass rates are difficult to interpret for many reasons ... By themselves, passing scores on licensure tests do not provide adequate information on which to judge the quality of teacher education programs ... The federal government should not use passing rates on initial licensure tests as the sole basis for comparing states and teacher education programs or for withholding funds, imposing other sanctions, or rewarding teacher education programs.

(Mitchell et al., 2001, pp. 170–171)

There are Real Problems and How Not to Fix Them

It is clear from both analyses initiated within the teacher education community (e.g., Wilson & Youngs, 2005), from critics of education schools such as Crowe (2010) and from impartial scientific panels convened by the National Research Council (Mitchell et al., 2001; NRC, 2010), that there are real problems with the teacher education accountability system in the U.S. that need to be addressed, including uneven standards for teachers and programs, different accountability rules for different kinds of programs, and the need to include a high quality measure of teaching effectiveness in both the initial licensing process and the assessment of the quality of teacher education programs. No one has argued that the current accountability system for teacher education programs is sufficient and does not need to be improved.

One approach that has become common in recent years is for vocal advocates of deregulation in teacher education and critics of education schools to proclaim themselves as "non-partisan" and issue their own evaluations of teacher education programs and reports on teacher education issues. There is no better example of this than the reports on teacher education programs that have been issued by the National Council on Teacher Quality (NCTQ) in the past few years.

Without any effort to submit either frameworks or "findings" to genuinely impartial peer review, the NCTQ proclaims:

Both program approval standards set by states and accreditation standards set by private organizations provide no indication of the quality of one institution's preparation relative to another ... Unfortunately this leaves consumers, aspiring teachers and schools who hire teachers in the dark ... As a non-partisan research and advisory organization committed to ensuring that every child has an effective teacher, NCTQ is stepping into this vacuum to help consumers distinguish between good, bad and mediocre education schools. We do so by setting the bar higher than it has been set traditionally.

(NCTQ, 2010)[19]

This allegedly non-partisan body which is not recognized by the federal government or any professional association as an accrediting body has issued its own set of standards for defining a high quality teacher education program[20] and has begun to go from state to state in applying its frameworks and issuing reports on the quality of different teacher education programs and on teacher education programs nationally in particular subject areas such as reading and mathematics (Walsh, Glaser, & Wilcox, 2006). The most recent report on teacher education programs in specific states is focused on Texas (NCTQ, 2010). This group, just like other groups, has the right to make its arguments about what makes a good teacher education program. There are two fundamental problems, though, with the current strategy of the NCTQ.

First, although the standards used to evaluate teacher education programs are described as representing consensus thinking from an impartial group, as illustrated in the quote below, the members of the group who developed the standards include some of the most outspoken critics of education schools and advocates of the deregulation of K-12 and teacher education, such as Chester Finn, Michael Podgursky, Frederick Hess, Michael Feinberg. Kate Walsh, and Michelle Rhee (Fordham Foundation, 1999; Hess, 2001; Walsh, 2004). This is hardly a non-partisan group.[21]

> The standards were developed over 5 years of study and are the result of contributions made by leading thinkers and practitioners from not just all over the nation but all over the world. To the extent that we can, we look to the practices of higher performing nations; where relevant the practices of other professions, and the best consensus thinking.
>
> (NCTQ, 2010)

Second, for a group that has focused so much on so called "scientific approaches" to teaching reading and mathematics,[22] it is ironic that they have not submitted their work to scholarly venues where it can undergo rigorous peer review and critique. Both the recent AERA and National Research Council investigations of research on U.S. teacher education underwent various levels of peer review prior to and after the release of the reports (Cochran-Smith & Zeichner, 2005; National Research Council, 2010). Why is the NCTQ so reluctant to have its judgments of teacher education programs undergo rigorous and impartial peer review in the most highly regarded journals?[23] Their strategy has been to go directly to the media with their so-called scientific reports. The media in turn print accounts about the conclusions of the NCTQ reports implying that they have undergone the usual scientific peer review. Sometimes, as in an article in the *Houston Chronicle* (Mellon, 2010) about the report on Texas, the media quote a few teacher educators who question the NCTQ methodology, but this is "balanced" by quotes from Texas superintendents who endorse the report. The NCTQ website includes a place for superintendents to indicate their support for its work.

This deceptive process of evaluating teacher education programs in an allegedly objective way is driven by a political agenda to deregulate teacher education rather than by any sense of scientific rigor. With regard to both the VAA bandwagon and the uncritical acceptance of the NCTQ reports, the media[24] have acted irresponsibly by not publicly discussing existing debates about issues like VAA or indicating whether or not a report has undergone genuine scientific peer review. To be fair, advocates of education schools have sometimes behaved in the same unscientific manner by issuing reports that are not subjected to rigorous peer review, and individuals from both the professionalization and deregulation camps have oversimplified and distorted the positions of their critics to some extent (Wilson & Tamir, 2008), and the media reports have sometimes included brief quotes from educators with different viewpoints.

Rising Above the Bickering

A strong public school system is an essential element of our democratic society, and given what we know about the importance of teachers to the quality of educational outcomes (National Academy of Education. 2009), preparing good teachers for everyone's children who attend our public schools is an extremely important activity that should be above partisan bickering. What our country needs is an accountability system in teacher education that is the result of open and reasoned discussion and debate of different positions on goals and the means to achieve them, and genuine peer review of research findings and policy recommendations. The cost-benefit framework suggested by Levin (1980), which calls for careful consideration of the social benefits and costs associated with particular elements of a teacher education accountability system, would be a useful way to structure this analysis.

The uncritical acceptance by the media of the pronouncements of *any* group on teacher education accountability interferes with the important goal of strengthening our teacher education accountability system. It is urgent for the U.S. Department of Education to commission the impartial evaluation of teacher education accountability called for in the National Research Council (2010) assessment of teacher education in the U.S. and for the Department to insist that the commonly acceptable standards for research including peer review be followed in the allocation of funds to support particular practices and policies in teacher education.

It has become very clear that public policies do not follow in any field in a linear way from research findings and that research can never dictate the specifics of particular policies in education or any other field (e.g., Kitson, Harvey, & McCormick, 1998; Stevens, 2007). Advocating for more reasoned and careful examination of the findings of particular inquiries related to accountability in teacher education and the methodologies that were used to produce them does not suggest that these inquiries will be able to translate directly into specific policies. There are legitimate differences that need to be negotiated in views about fundamental aspects of teaching and teacher education related to the purposes of public education, the role of teachers, how student learning can be measured, and so on that will never be able to be resolved through research alone, even if it is of high quality (Cochran-Smith & Zeichner, 2005). This open debate and discussion about the goals and processes of public education and teacher education are fundamental benefits of living in a democratic society, and we should insist that our government agencies and the media support rather than short-circuit this process.

Former Illinois state superintendent of education Joseph Cronin, in his analysis of the history of state regulation in teacher education in the U.S., warns policymakers about the dangers of supposedly simple solutions[25] to problems of teacher education accountability and of avoiding the kind of reasoned discussion and debate that is needed:

> Most of all, legislators and study commission members should remember that any change not only may fail to solve the specific problem but may in fact create new problems not anticipated at present ... Remember the immortal words of H.L. Mencken, those of you who would reform teacher education: For every complicated problem there is a simple solution and it is usually wrong.
>
> (1983, p. 190)

Redesigning the accountability system for teacher education in the U.S. is a complex matter that requires all of us to rise above our own self-interest and to learn to work in more productive ways with those who hold positions different from our own.

Notes

1 I would like to thank the following people, in addition to the editors, for their helpful comments about earlier drafts: Michael Apple, Linda Darling-Hammond, Mary Diez, Penny Engel, Kerry Kretchmar, Katie Payne, Cap Peck, Sharon Robinson, Cathy Taylor, Sheila Valencia, and Pat Wasley.
2 These include program approval, testing of basic skills, content and professional knowledge, and various ways to assess the quality of teachers' teaching.
3 As Conant (1963) and Cronin (1983) have pointed out, the teaching profession through the National Commission on Teacher Education and Professional Standards (TEPS), which was an arm of the National Education Association, and the professional standards boards in some states has exerted varying degrees of influence on both teacher certification and teacher education program accountability. They have also pointed out that states have differed in the degree to which they have given higher education a role in determining and monitoring these processes.
4 Forty-three states have adopted or integrated criteria for assessing the quality of teacher preparation programs from voluntary national accreditation agencies (U.S. Department of Education, 2009).
5 There have always been some tests involved in getting an initial teaching license (e.g., Elsbree, 1939), but many of them were locally administered by school districts or county education officers.

6 These figures with regard to the numbers of states using particular kinds of certification tests come from the most recent report on teacher quality from the U.S. Secretary of Education (U.S. Department of Education, 2009).

7 The term "C/BTE" is being used in a general way here as it was back in the 1970s to describe a general approach to teacher education that focused on teacher candidates demonstrating mastery of a set of outcomes. In practice, programs ranged from those that focused on discrete bits of isolated aspects of teaching, while others focused on fewer and more integrated aspects of teaching based on a clear conceptual framework (Liston & Zeichner, 1991).

8 The data on the NASDTEC website on August 1 are from 2004.

9 See Wasley and McDiarmid (2004) for a discussion of a number of different ways to connect teacher education, teaching, and student learning.

10 The highly publicized Conant Report (1963) emphasized the improvement of the clinical component of teacher education as the most important thing that could be done to raise the quality of teacher education in the U.S. and singled out the lack of preparation and support for cooperating teachers as one of the weakest aspects of the system. For example, "cooperating teachers should have time freed to aid the student teachers; they should also have increased compensation in recognition of their added responsibility and talent" (p. 62).

11 http://aacte.org/index.php?/Programs/Teacher-Performance-Assessment-Consortium-TPAC/teacher-performance-assessment-consortium.html

12 Detailed information about the PACT can be found at http://www.pacttpa.org/_main/hub.php?pageName=Home, and information about the CCSSO and AACTE project to develop a nationally available performance assessment based on PACT can be found at http://aacte.org/index.php?/Programs/Teacher-Performance-Assessment-Consortium-TPAC/teacher-performance-assessment-consortium.html

13 See http://www.ets.org/praxis/institutions/praxisiii/

14 See http://www.wou.edu/education/worksample/twsm/

15 There have been recent efforts to streamline and focus national accreditation more on outcomes (www.ncate.org, www.teac.org).

16 NCATE is already involved in streamlining its system and strengthening its connection to P-12 student learning and its ability to support the continuous improvement of programs (Cibulka, 2009).

17 The current fees for the most widely used tests are: Praxis 1 ($130), Praxis II ($65–115), and a $50 registration fee. See http://www.ets.org/praxis/about/fees

18 One could argue that requiring candidates to pass a content test prior to their final student teaching or internship experience (which is a common practice) is a more ethical stance to take, given the effects that lack of minimal content knowledge could have on pupils, and the investment of time and money that candidates need to do a full-time clinical experience. I am grateful to Mary Diez for pointing this out to me.

19 Retrieved from www.nctq.org/p/response/evaluation_faq.jsp on August 7, 2010.

20 The NCTQ standards focus on admissions and exit requirements from programs and the preparation of teachers to teach reading and mathematics. They do not address the quality of performance of teacher candidates in classrooms (www.nctq.org).

21 On its website (nctq.org), its mission is stated as "to provide an alternative voice to existing organizations and build a case for a comprehensive reform agenda that would challenge the current structure and regulation of the profession." The problem is not that NCTQ has involved mostly supporters of a deregulation agenda and critics of education schools to advise them on the development of their standards. I am also not suggesting that some amount of deregulation in teacher education is a bad thing. The problem is that the NCTQ publicly describes its position as non-partisan, purely in service of consumers, and that it is not transparent about the political agenda that propels its work.

22 Assertions that are in conflict with the recent analyses of the National Research Council (2010).
23 See http://aacte.org/index.php?/Traditional-Media/Resources/aacte-members-respond-to-nctq-qresearchq-efforts.html for a series of letters from state professional teacher education associations and education school deans detailing some of the methodological and ethical concerns that exist about the NCTQ evaluations.
24 This includes some of the major national publications in education such as the *Chronicle of Higher Education* and *Education Week*.
25 Teacher warranties where teacher education programs guaranteed the quality of their graduates, and promised to remediate any deficiencies at no cost to school districts, is one of the allegedly simple solutions to teacher education program accountability that did not amount to much (Earley, 2000b).

References

Abramson, L. (2010). Study tries to track LA teachers. January 4 edition of *All things considered*. Retrieved January 10, 2010 from www.npr.org

Acheson, K. & Gall, M. (1980). *Techniques in the clinical supervision of teachers*. New York: Longman.

American Association of Colleges for Teacher Education (1974). *Achieving the potential of performance-based teacher education*. Washington, DC: Author.

American Association of Colleges for Teacher Education (2010, March). *The clinical preparation of teachers: A policy brief*. Washington, DC: Author.

Bales, B. (2006). Teacher education policies in the United States: The accountability shift since 1980. *Teaching and Teacher Education 22*, 395–407.

Berlak, A. (1010). Coming soon to your favorite credential program: National exit exams. *Rethinking schools 25*(1). Retrieved August 30, 2010 from www.rethinking schools.org

Berry, B. (2010, May). *Strengthening state teacher licensure standards to achieve teaching effectiveness*. Washington, DC: Partnership for Teacher Quality. Retrieved June 24, 2010 from www.aacte.org

Boyd, D. et al. (2008). Surveying the landscape of teacher education in New York City: Constrained variation and the challenge of innovation. *Educational Evaluation and Policy Analysis 30*(4), 319–343.

Braun, H. (2005). *Using student progress to evaluate teachers: A primer on value-added models*. Princeton, NJ: Educational Testing Service.

Brennan, J. (1999). *Study finds teacher licensure tests to be mostly high school level: If this is all we expect teachers to know why send them to college?* Washington, DC: The Education Trust.

Center for American Progress. (2010). Better teachers, better students: Event proposes ways to measure the effectiveness of teacher training programs. Retrieved August 3, 2010 from www.americanprogress.org

Chung Wei, R. & Pecheone, R. (2010). Assessment for learning in preservice teacher education. In M. Kennedy (Ed.), *Teacher assessment and the quest for teacher quality* (pp. 69–132). San Francisco: Jossey-Bass.

Cibulka, J. (2009). The redesign of accreditation to inform simultaneous transformation of educator preparation and P-12 schools. *Quality Teaching 18*(2). Retrieved August 1, 2010 from www.ncate.org

Clarke, S. (1969). The story of elementary teacher education models. *Journal of Teacher Education 20*(3), 283–293.

Cochran-Smith, M. & Zeichner, K. (2005) (Eds.). *Studying teacher education*. New York: Routledge.

Conant, J. (1963). *The education of American teachers*. New York: McGraw Hill.

Cronin, J.M. (1983). State regulation of teacher preparation. In L. Shulman & G. Sykes (Eds.), *Handbook of teaching and policy* (pp. 171–191). New York: Longman.

Crowe, E. (2010, July). *Measuring what matters: A stronger accountability model for teacher education*. Washington, DC: Center for American Progress.

Danielson, C. (1996). *Enhancing professional practice: A framework for teaching*. Alexandria, VA: Association of Supervision and Curriculum Development.

Darling-Hammond, L. (2009). Recognizing and enhancing teaching effectiveness. *International Journal of Educational and Psychological Assessment 3*, 1–24.

Darling-Hammond, L. (in press). *Evaluating teacher effectiveness: How teacher performance assessments can measure and improve teaching*. Washington, DC: Center for American Progress.

Darling-Hammond, L. & Baratz-Snowden, J. (Eds.). (2005) *A good teacher in every classroom*. San Francisco: Jossey-Bass.

Darling-Hammond, L. & Chung Wei, R. (2009). Teacher preparation and teacher learning: The changing policy landscape. In G. Sykes, B. Schneider, & D. Plank (Eds.), *Handbook of education policy research* (pp. 613–636). New York: Routledge.

Darling-Hammond, L., Pacheco, A., Michelli, N., LePage, P., Hammerness, K. & Youngs, P. (2005). Implementing curriculum renewal in teacher education: Managing organizational and policy change. In L. Darling-Hammond & J. Bransford (Eds.), *Preparing teachers for a changing world* (pp. 442–479). San Francisco: Jossey-Bass.

Delandshere, G. & Petrosky, A. (2010). The use of portfolios in preservice teacher education. In M. Kennedy (Ed.), *Teacher assessment and the quest for teacher quality* (pp. 9–42). San Francisco: Jossey-Bass.

Diez, M. (2010). It is complicated: Unpacking the flow of teacher education's impact on student learning. *Journal of Teacher Education 61*(5), 441–450.

Diez, M. & Haas, J. (1997). No more piecemeal reform: Using performance assessment to rethink teacher education. *Action in Teacher Education 19*(2), 17–26.

Duncan, A. (2009, October). *Teacher preparation: Reforming the uncertain profession*. Address given by Secretary of Education Arne Duncan at Teachers College, Columbia University.

Earley, P. (2000a) Finding the culprit: Federal policy and teacher education. *Educational Policy 14*(1), 25–39.

Earley, P. (2000b, February). *Guaranteeing the quality of future educators: Report on a survey of teacher warranty programs*. Washington, DC: American Association of Colleges for Teacher Education.

Elsbree, W. (1939). *The American teacher*. New York. American Book Co.

Fordham Foundation (1999). *Better teachers and how to get more of them*. Dayton, OH: Author.

Fraser, J. (2007). *Preparing America's teachers: A history*. New York: Teachers College Press.

Gage, N. & Winne, P. (1975). Performance-based teacher education. In K. Ryan (Ed.), *Teacher education* (pp. 146–172). Chicago: University of Chicago Press.

Gitomer, D., Latham, A.S. & Ziomek, R. (1999). *The academic quality of prospective teachers: The impact of admissions and licensure testing*. Princeton, NJ: Educational Testing Service.

Glenn, D. (2010). Education schools are scrutinized for graduates' success as teachers. *Chronicle of Higher Education*. Retrieved on August 5, 2010 from http://chronicle.com

Goldhaber, D. (2007). Everyone's doing it, but does it tell us about teacher effectiveness? *Journal of Human Resources 52*(4), 765–794.

Goldhaber, D. (2010). Licensure tests: Their use and value for increasing teacher quality. In M. Kennedy (Ed.), *Teacher assessment and the quest for teacher quality* (pp. 133–162). San Francisco: Jossey-Bass.

Goldhammer, R. (1969). *Clinical supervision: Special methods for the supervision of teachers*. New York: Holt, Rinehart & Winston.

Hamel, F. & Merz, C. (2005). Reforming accountability: A preservice program wrestles with mandated reform. *Journal of Teacher Education 56*(2), 157–167.

Harris, D. & McCaffrey, D. (2010). Value added: Assessing teachers' contributions to student achievement. In M. Kennedy (Ed.), *Teacher assessment and the quest for teacher quality* (pp. 251–283). San Francisco: Jossey-Bass.

Heath, R.W. & Nielson, M. (1974). The research base for performance-based teacher education. *Review of Educational Research 44*(4), 463–484.

Hess, F. (2001). *Tear down the wall: The case for a radical overhaul of teacher certification*. Washington, DC: Progressive Policy Institute.

Honowar, V. (2007). Gains seen in retooled teacher education. *Education Week*. Retrieved October 27, 2007 from www.edweek.org

Houston, W.R. & Howsam, R. (1972). *Competency-based teacher education*. Chicago: Science Research Associates.

Imig, D. & Imig, S. (2008). From traditional certification to competitive certification. In M. Cochran-Smith, S. Feiman-Nemser, & D.J. McIntyre (Eds.), *The handbook of research on teacher education*, 3rd edition (pp. 886–907). New York: Routledge.

Johnson, D., Johnson, B., Farenga, S., & Ness, D. (2005). *Trivializing teacher education: The accreditation squeeze*. Lanham, MD: Roman Littlefield.

Kelderman, E. (2010). Teacher-education programs are unaccountable and undemanding report says/ *Chronicle of Higher Education*, July 29. Retrieved July 30, 2010 from http://chronicle.com

Kitson, A., Harvey, G., & McCormick, B. (1998). Enabling the implementation of evidence-based practice: A conceptual framework. *Quality on healthcare, 7*, 149–158.

Kornfeld, J., Grady, K., Marker, P., & Ruddell, M. (2007). Caught in the current: A self study of state-mandated compliance in a teacher education program. *Teachers College Record 109*(8), 1902–1930.

Levin, H.M. (1980). Teacher certification and the economics of information. *Education Evaluation & Policy Analysis 2*(4), 5–18.

Liston, D. & Zeichner, K. (1991). *Teacher education and the social conditions of schooling*. New York: Routledge.

Lyall, K. & Sell, K. (2006). *The true genius of America at risk: Are we losing our public universities to de facto privatization?* Westport, CT: Praeger.

Matus, R. (2009). State rates teacher prep programs. *St. Petersburg Times*, November 19. Retrieved January 4, 2010 from www.tampabay.com

McCaffrey, D.M., Lockwood, J.R., Mariano, L., & Setodji, C. (2005). Challenges for Value-added assessment of teacher effects. In R. Lissitz (Ed.), *Value-added models in education: Theory and applications* (pp. 111–144). Maple Grove, MN: JAM Press.

McConnery, A.A., Schalock, M.D., & Schalock, H.D. (1998). Focusing improvement and quality assurance: Work samples as authentic performance measures of prospective teachers' effectiveness. *Journal of Personnel Evaluation in Education 11*, 343–363.

Mellon, E. (2010). Colleges slammed over teacher preparation. *Houston Chronicle*, April 28. Retrieved August 6, 2010 from texas_houstonchronicle_article_apr282010.pdf

Mitchell, K., Robinson, D., Plake, B., & Knowles, K. (2001). *Testing teacher candidates: The role of licensure tests in improving teacher quality*. Washington, DC: National Academy Press.

National Academy of Education (2009). *Teacher quality*. Retrieved January 2, 2010 from www.naeeducation.org NASDTEC (2010). Retrieved August 22, 2010 from www.nasdtec.org

National Council on Teacher Quality (2010, April). Ed school essentials: Evaluating the fundamentals of teacher training programs in Texas. Retrieved on August 6, 2010 from http://www.nctq.org/edschoolreports/texas

National Research Council (2010, April). *Preparing teachers: Building evidence for sound policy.* Washington, DC: National Academies Press.

Neville, K.S., Sherman, R.H., & Cohen, C.E. (2005). *Preparing and training professionals: Comparing education to six other fields.* New York City: The Finance Project. Retrieved January 6, 2007 from www.financeproject.org

Newton, S., Walker, L., & Darling-Hammond, L. (2010). *Predictive validity of the Performance Assessment for California Teachers.* Stanford University: Stanford Center for Opportunity Policy in Education.

Noell, G.H. & Burns, J.L. (2006). Value-added assessment of teacher preparation: An illustration of emerging technology. *Journal of Teacher Education 57*(1), 37–50.

Pecheone, R. & Chung, R. (2006). Evidence in teacher education: The performance Assessment for California Teachers. *Journal of Teacher Education 57*(1), 22–36.

Peck, C., Gallucci, C., & Sloan, T. (2010). Negotiating implementation of high-stakes performance assessment policies in teacher education: From compliance to inquiry. *Journal of Teacher Education 61*(5), 451–460.

Pianta, R. & Hamre, B. (2009). Conceptualization, measurement, and improvement of classroom processes: Standardized observation can leverage capacity. *Educational Researcher 38*(2), 109–119.

Porter, A., Youngs, P., & Odden, A. (2001). Advances in teacher assessments and their uses. In V. Richardson (Ed.), *Handbook of Research on Teaching,* 4th edition (pp. 259–297). Washington, DC: American Educational Research Association.

Prestine, N. (1989). The struggle for the control of teacher education: A case study. *Educational Evaluation & Policy Analysis 11*(3), 285–300.

Rennert-Ariev, P. (2008). The hidden curriculum of performance-based teacher education. *Teachers College Record, 110*(1), 105–138.

Rothstein, J. (2010). Teacher quality in educational production: Tracking, decay, and student achievement. *Quarterly Journal of Economics 123*(1), 175–214.

Simon, A. & Boyer, G. (1974). *Mirrors for behavior: An anthology of classroom observation instruments,* vol. 3. Philadelphia: Research for Better Schools.

Stevens, A. (2007). Survival of the ideas that fit: An evolutionary analogy for the use of evidence in policy. *Social Policy and Society 6*(1), 25–35.

Sykes, G. (2009, March). *Fifty years of federal teacher policy: An appraisal.* Washington, DC: Center on Education Policy.

Sykes, G. & Dibner, K. (March, 2009). *Fifty years of federal teacher policy: An appraisal.* Washington, DC: Center on Education Policy.

U.S. Department of Education (2009, November). *The secretary's sixth annual report on teacher quality.* Washington, DC: Office of Postsecondary Education.

Valli, L. & Rennert-Ariev, P. (2002). New standards and assessments? Curriculum transformation in teacher education. *Journal of Curriculum Studies 34*(2), 201–206.

Villegas, A.M. & Davis, D.E. (2008). Preparing teachers of color to confront racial/ethnic disparities in educational outcomes. In M. Cochran-Smith, S. Feiman-Nemser, & D.J. McIntyre (Eds.), *Handbook of research on teacher education,* 3rd edition (pp. 583–605). New York: Routledge.

Walsh, K. (2004). A candidate-centered model for teacher preparation and licensure. In F. Hess, A. Rotherham, & K. Walsh (Eds.), *A qualified teacher in every classroom?* (pp. 223–254). Cambridge, MA: Harvard Education Press.

Walsh, K., Glaser, D., & Dunne Wilcox, D. (2006, May). *What education schools aren't teaching about reading and what elementary teachers aren't learning.* Washington, DC: National Council on Teacher Quality.

Wang, A.H., Coleman, A.B., Coley, R.J., & Phelps, R.P. (2003, May). *Preparing teachers around the world.* Princeton, NJ: Educational Testing Service.

Wasley, P. & McDiarmid, G.W. (2004, June). *Connecting the assessment of new teachers to student learning and to teacher preparation.* Prepared for the National Commission on Teacher and America's Future Summit on High Quality Teacher Preparation, June 28–30, Austin, TX. Retrieved August 4, 2009 from http://www.nctaf.org/resources/events/2004_summit-1/documents/Wasley-McDiarmid_Final_-_NCTAF.pdf

Wilkerson, J.R. & Lang, W.S. (2003). Portfolios, The pied piper of teacher certification assessments: Legal and psychometric issues. *Education Policy Analysis Archives* 11(45). Retrieved February 12, 2004 from http://epaa.asu.edu/eppa

Wilson, S. (2009). Measuring teacher quality for professional entry. In D.H. Gitomer (Ed.). *Measurement issues and assessment for teaching quality.* (pp. 8–29). Thousand Oaks, CA: Sage.

Wilson, S., Floden, R., & Ferrini-Mundi, J. (2001). *Teacher preparation research: Current knowledge, gaps, and recommendations.* Seattle: Center for the Study of Teaching and Policy, University of Washington, College of Education.

Wilson, S. & Tamir, E. (2008). The evolving field of teacher education. In M. Cochran-Smith, S. Feiman-Nemser, & D.J. McIntyre (Eds.), *Handbook of research on teacher education,* 2nd edition (pp. 908–935). New York: Routledge.

Wilson, S. & Youngs, P. (2005). Research on accountability processes in teacher education. In M. Cochran-Smith & K. Zeichner (Eds.), *Studying teacher education* (pp. 645–736). New York: Routledge.

Wisconsin Journal of Education (1863). Examination of teachers. *Wisconsin Journal of Education* 7, 19.

Zeichner, K. (2005). Learning from experience with performance-based teacher education. In F. Peterman (Ed), *Designing performance assessment systems for urban teacher education* (pp. 3–19). New York: Routledge.

Zeichner, K. (2006). Reflections of a university-based teacher educator on the future of college and university-based teacher education. *Journal of Teacher Education* 57(3), 326–340.

EDITORS' COMMENTARY

This chapter illustrates the complex interaction of institutional, state, and federal teacher education policies. Although there is general support for some form of professional accreditation among teacher educators, opinions in the policy community are mixed. Some state policymakers think accreditation by NCATE or TEAC is important enough to be mandatory, whereas others dismiss it as not relevant at all. The latter perspective is found in Chapter 4, the policy case study of Florida. As Zeichner notes, the link between most accountability measures is tenuous, but that has not stopped federal decision makers from imposing certain of these measures on the teacher education system, specifically through requirements to receive federal funds. One might wonder, then, why federal policymakers have not required professional accreditation, such as NCATE or TEAC, for all teacher preparation programs—those based in institutions of higher education as well as alternative route programs found elsewhere.

The answer is twofold. Many federal decision makers, both in Congress and in the U.S. Department of Education, are skeptical of the ability of those within teacher education to objectively judge their peers. Even if there was a direct correlation between professional accreditation and teacher quality, it is unlikely decision makers' skepticism would be diminished. In addition, it is important to note that there are many education lobbying groups offering perspectives on teacher education and other issues. Among these groups are national associations that represent college and university presidents. They include the American Council on Education, the National Association of State Colleges and Universities, and the National Association of Independent Colleges and Universities. Teacher education programs housed in colleges and universities are represented by one or more of these three organizations and they share a common tenet that federal intervention in higher education issues should be minimal. Moreover, their members—the college and university presidents—are wary of needing to respond to the demands of multiple professional accreditation bodies due to the cost of compliance and the demands placed on institutions. The case study of Florida is illustrative of this. Although college presidents were part of a blue ribbon commission on teacher education, they apparently did not defend a requirement that their education units retain NCATE accreditation. The organizations that represent college presidents have opposed federal mandates for teacher preparation accreditation by NCATE or TEAC in the past and it is likely this position will not change. The complex web of connections between federal expectations, interest group positions, revolving state requirements, and what is important to the teacher education community illustrates the sense-making and co-construction perspectives put forward by Datnow and Park (and presented in Chapter 1). As policies are put into place, there are nuanced interactions between policy actors and their decisions, and these interactions are constantly being influenced by the context in which they are implemented.

An important theme in Zeichner's work is that, time after time, policymakers jump to a decision whether there is empirical evidence to support that decision or not. He cites the use of value-added assessments as a contemporary example. Sykes documents many other instances when federal policies not only have no supporting evidence but were enacted in the face of contrary research (see Chapter 1). Unfortunately, if state and federal decision makers do not trust educators and teacher educators in particular, it is not surprising that they would dismiss scholarly findings conducted by them. Another aspect of the evidence and policy disconnect is the use of language. Virtually every major policy initiative since the late 1990s has referenced teacher quality: high quality is good, and poor quality is not. But drawing meaning from these words is problematic. In the Higher Education Act, a qualified teacher is one who passes the state licensing exam, but passing an exam alone does not guarantee successful teaching in all situations. The

No Child Left Behind Act defines characteristics of a highly qualified teacher which includes an academic major or minor in the teaching field. But, as Zeichner observes, how can a college major or minor guarantee an individual will perform well in the classroom? Deborah Stone (see Chapter 1) discusses the problem of multiple understandings of terms as contributors to the paradoxical nature of creating and implementing policies, and in the realm of teacher education policy this clearly is the case.

In his essay in Chapter 2, Hess joins Zeichner in questioning the utility of the current generation of value-added assessments as a means of deciding who is a good teacher and who is not. Zeichner argues that the most promising mechanism for deciding which teachers are effective is to observe them over time as they engage in their work. He acknowledges this is a time-consuming and expensive form of evaluation. This is true. What also is true is that it seems unlikely that a society unwilling to pay teachers a salary commensurate with the challenges of their work will be willing to support investments in costly evaluation mechanisms.

Zeichner offers the recommendation that comprehensive evaluation systems, both of teacher education programs and teachers themselves, be built through partnerships and structures such as professional standards boards. Chapters 6 and 7 present longitudinal policy case studies of partnerships, but the outcomes were not the same. In New Jersey the establishment and implementation of a K-16 partnership between an institution and school district is described from the points of view of a dean, superintendent, and teacher. All attest to the success of the venture. Weisenbach's description of the rise and demise of an independent professional standards and licensure board in Indiana suggests a less optimistic picture. The Indiana standards board, which included representatives from the K-12 and higher education community, seemed to enjoy initial success until key actors and the policy context changed. In New Jersey, there was continuity of leadership in the university during the decade in which the partnership was developed and implemented. The lesson here may be that K-16 partnerships, whether they are a quasi-governmental unit like a practices board or an arrangement between a school district and a university, need to be robust enough to withstand shifts in context and changes in leadership.

Discussion Questions

1. If one of the policy dilemmas is agreeing on common understandings and definitions what terms could be used to describe a teacher candidate who is ready to enter the classroom but whose experience in multiple situations has not been observed? Are there terms to describe teachers observed as effective in multiple settings? How would educators and policymakers reach agreement on these terms and how to quantify them?
2. Given the reality that leaders change jobs and every election cycle brings the opportunity for a new set of policymakers to replace existing ones, how can

K–16 educators protect successful partnerships from disruption by contextual changes?

3. If traditional teacher education programs are constrained by state policy and may lack support from college or university presidents, is there a future for collegiate-based teacher preparation?

4. Zeichner provides some critiques of policy and the analysis of policy, suggesting that some—Crowe, for example—selectively and inappropriately use other professions as examples in their criticism of teacher education. Do these arguments seem likely to influence policymakers and others?

6

SHALLOW ROOTS

The Effect of Leaders and Leadership on State Policy

E. Lynne Weisenbach

The vision was lofty—a fully aligned system of teacher preparation, induction, and professional development. Teacher preparation was to be based on standards, with authentic clinical assessments of teacher candidates based on those standards. Induction of new teachers would include a strong mentoring component, and standards for teachers who would serve as mentors were to be designed and an approval process for mentor training implemented. A tiered licensure system was to be built that would parallel the preparation, induction and professional development system. Built off the work Linda Darling-Hammond (1997), the so-called "three-legged stool" of accreditation, licensing, and certification was to lead to a dramatically improved system of teacher preparation and professional development in Indiana (p. 63). This narrative will provide the history of what many thought would be a powerful and productive reform—one that, in the end, had shallow roots.

Governance of teacher preparation and licensing in the United States is a function of the states. Many states have semi-independent professional standards and practices boards, which make recommendations to the state board of education regarding standards for licensure of K-12 teachers/administrators, standards for preparation programs, and monitoring of ethics/professional practice. A minority of states have autonomous, independent professional standards and practices boards, which are established by state statute and are accountable directly to the state legislature. These autonomous boards address the foregoing criteria; however, they report to the legislature and not to a state board of education. Still other states have variations on these two models.

Scannell and Wain note that:

> The authority for setting standards for teacher preparation and licensing and for governing the practice of the profession has historically resided with

other bodies, such as state legislatures, state boards of education, or state departments of education. Educators have then been held accountable for standards developed and enforced by others, who often view teaching as semiskilled work. In many cases, those who set licensing policies are still ambivalent about whether there is a knowledge base for teaching or whether teacher preparation makes a difference, even though research clearly demonstrates that those who complete such preparation are more effective teachers.

(1996, p. 211)

Determination about the governance of teacher preparation and licensing is not permanent in any given state. For example, during the late 1990s, models of professional governance evolved and, in many cases, changed with more control ceded to state boards, legislatures, and even governors. A variety of factors come into play, primarily philosophy and politics. Consequently, statutes, administrative codes, or policies governing boards of teaching can—and do—change.

Creation and Early Implementation

Sitting in the "heartland" of the United States, Indiana is home to the famed Indianapolis Motor Speedway. "Speed and racing" provide an appropriate metaphor for the pace at which reform of teacher preparation, induction, and professional development was conceived and implemented in the Hoosier state in the early 1990s.

Driven by awareness that teacher quality mattered, and a need for dramatic change, the Indiana Professional Standards Board (IPBS) was created by the Indiana Legislature in 1992, following a study commission recommendation that called for the creation of an autonomous teacher licensing and accreditation board. This recommendation and subsequent actions by the Indiana legislature were based on a promise of self-governance, which is taken for granted by many professions, yet, at the time, and still today, is the exception in U.S. education. The board was to have 16 members, 15 of whom would be appointed by the Governor. The profession was to be well represented, and was to include teachers and administrators. The 16th member was the state superintendent of education. In Indiana, as in many states, the state superintendent was an elected position. Two years later, there was a recommendation to create a separate autonomous governing board for school and district administrators. The request was prompted in part because of the preponderance of teachers on the new board, which resulted, in effect, in teachers governing the preparation and licensure of school administrators. In a compromise move, three new members, including another administrator, a special education director, and a representative from the business community, were added to the IPSB. From that point forward, the board had 19 members, with all but one, the state superintendent, appointed by the Governor. Nine were teachers; three representatives were from higher education; there were two building-level administrators, one district or area superintendent, and one special education director. There was also one representative of business, one school board member, and the state superintendent of education. The new board's charge was to set standards for preparing and licensing education professionals, develop an assessment system to determine if standards were met, and to do a complete revamp of the licensure system for educators so that it would link directly to standards.

The budget and associated funds to support licensing and teacher preparation within the Department of Education were transferred immediately to the new board, which, in turn, assumed full monetary control. A staff member who had overseen these functions under the state superintendent at the Department was appointed director, and many of the staff transferred to the new unit. Consequently, there was a capable and knowledgeable base on which to create the new entity.

The evolution of the board from a group of committed individuals drawn from many different parts of the state and across different levels of the system to a highly functioning, visionary board was not immediate. Many on the board did not have prior board experience, there were those who asked why the board was necessary, and, most importantly, there was not a shared vision for the board or for teacher preparation/licensing. To that end, and with support from the Lilly Endowment, Inc., the board immediately began a review of current literature about teacher preparation and licensing and sought the counsel of leading national experts. These included, but were not limited to, highly regarded leaders drawn from the American Association of Colleges for Teacher Education (AACTE) and the National Council for Accreditation of Teacher Education (NCATE).

Within a year, the new board had become a cohesive group, was well versed in national and state issues in teacher preparation and licensing, and had created a mission statement that included their role in "establishing and maintaining rigorous, achievable standards for educators, beginning with preservice [preparation] and continuing throughout a teacher's professional career" (Indiana Department of Education [IDOE], n.d., "Fulfilling the mission" section, para. 2).

Leadership

The goal of the Indiana Professional Standards Board was clear, understandable, and seemed attainable: design and implement a system for preparing and licensing teachers that would be based on demonstrated performance. At that time, "teacher performance" was regarded as demonstrating practices consistent with good teaching. Although it sounded simple and straightforward, attaining the goal would prove to be complicated, and would affect the status quo in ways few could have imagined. Among other things, there were no metrics, no longitudinal data systems, and minimal alignment of standards between the K–12 and postsecondary education systems. As a first step, in 1993, the IPSB began a review of the 19-year-old rules and regulations for teacher preparation and licensure. The pros and cons of these regulations were described and analyzed. After a thorough review, members of the new board agreed that a complete overhaul of the existing rules and regulations was needed.

In 1994, two years after the creation of the IPSB, the founding director announced his retirement. Under his leadership, the board had become unified and strong. At this juncture, they felt that Indiana needed someone with national perspective to lead the important work that was ahead. They recruited a Senior Director with experience at the American Association of Colleges for Teacher Education (AACTE), who had a clear passion to drive the board's vision. With the board, she designed a plan for creating and implementing sets of developmental and content standards that were tailored for Indiana, and also aligned to concurrent national initiatives, including the standards for practicing professionals from the National Board for Professional Teaching Standards (NBPTS). The new director served effectively as the public face of the IPSB, was skilled at working with a range of constituency groups, and was an effective communicator.

The Standards: Plan and Process

With visionary leadership in place, the IPSB acted quickly in 1994 and agreed to implement a performance-based system, linked to the ten Interstate New Teacher Assessment and Support Consortium (INTASC) standards. Created in 1987, INTASC was a consortium of state education agencies and national organizations dedicated to the reform of the preparation, licensing, and ongoing professional development of teachers. Faculty members from many institutions of higher education from across the country were engaged in the creation of the INTASC standards. Faculty members and K–12 teachers from Indiana were actively engaged in that effort, and consequently had a strong understanding of, and commitment to, the new standards.

The inclusionary nature of the standards setting process at the national level was replicated, with modification, by Indiana. The new director of the IPSB quickly proved that she was adept at building a strong, inclusionary infrastructure. Two initial advisory groups were formed to create/recommend standards for Board approval: one for teachers of mathematics, early childhood through young adult levels, and the other for early adolescent levels, regardless of content level. The advisory groups were created through an "open call" system to serve and were drawn from a broad base of faculty across the K–16 spectrum. The teachers, teacher educators, and university content area faculty selected to work collaboratively on the standards were assisted by national advisors. The national advisors provided important guidance and direction. It is significant that many faculty appointed to the groups also had ties to national professional organizations. In other cases, many members of the groups subsequently became affiliated with national organizations, based on the work and their contact with the national advisors. In other words, the advisory groups themselves became learning communities and advocates for a new system across the state. The structure of appointing, staffing, and utilizing these two groups served as a framework for the subsequent design of the remaining developmental and content standards, including the sciences and social studies.

One year later, the overall process design was complete and a state-wide call went out for the remaining 14 advisory groups. Of the 14, three were for developmental levels, and eleven were for content areas. The content area standards cut across developmental areas, and the developmental standards cut across the content areas. The intent was for overall alignment of both content and developmental areas. Although a crosswalk document was created, confusion existed over the relationship between the two. Over 200 people were involved in the creation of the new standards. With a mix of private (primarily from the Lilly Endowment) and state funding, those involved were compensated for travel for meetings; however, nearly all served on a voluntary basis. They became well versed in the INTASC standards and about the expectation that all standards would include components regarding the knowledge, dispositions, and performances expected of candidates or practicing teachers. By 1998, the creation of a set of 17 standards to be used for the initial licensing of beginning teachers in Indiana was complete.

The System: A House Under Construction

Concurrent with the standards development process, in 1996, the Board turned its attention to implementation of the standards through the design of a new preparation and licensing system. A widely used analogy was to compare the new system to a three-story house:

- The first floor was the standards that included knowledge, dispositions and performances.
- The second floor was a set of well-crafted assessments.
- The third floor was a parallel licensing configuration.
- The roof included the statutes, rules, and regulations necessary for implementation.

Additionally, each floor had three rooms: preparation, licensure, and professional development.

As the 17 sets of standards were being developed in 1997, voluntary advisory groups concurrently began work on the second floor, assessments that were to be aligned to the standards. At the national level, institutions such as Alverno College (Wisconsin) were on the leading edge of developing institutional systems based on student performance—not simply courses taken and grades earned. The Indiana Association of Colleges for Teacher Education (IACTE), an affiliate of the American Association of Colleges for Teacher Education (AACTE), was a strong partner with IPSB and provided professional development for campus leaders, especially the deans of colleges of education, regarding performance-based assessments. The engagement of IACTE served as an important means of engaging both the independent and public sectors in higher education in the state. Leadership of IACTE was designed in a way that all voices including those from large and small institutions, were at the table. Institutional differences were recognized and valued due to the open dialogue and shared challenges in building outcomes-based assessments and overall assessment systems. Additionally, there was ongoing dialogue and sharing between AACTE and IACTE, which was important because it ensured Indiana's access to new national developments, and also ensured that Indiana's "race forward" was leveraged nationally for maximum impact. Because of the composition of the standards advisory groups, faculty and teachers from across the state were already connected and collaborating. These networks proved to be valuable in ensuring K-16 collaboration and alignment.

By 2000, 40 institutions of higher education with teacher preparation programs, led by the deans of colleges of education with support from IACTE, were designing elaborate institutional unit assessment systems based on the standards. "Unit" was the term to designate the campus-level entity responsible for teacher preparation. It was usually, although not always, a college of education. These designs generally included a scaffolded approach to assessing a candidate's readiness to teach based on assessments that included:

- assessments and high stakes decisions at the entrance to the teacher education program,
- assessments at the midpoint of a program, including some demonstrations in classrooms, and
- assessments at the completion of the program, sometimes termed "exit" assessments. Successful completion was generally necessary for program completion and/or recommendation for licensure.

The systems were to measure aspects of the standards in multiple ways, i.e., knowledge that the candidate should possess; ability to apply the knowledge in the classroom; dispositions and behaviors associated with successful teaching; and the ability to have positive effects on student learning. Additionally, the system was to have rubrics that defined levels of performance, the ability to summarize candidate performance, and provide credible and useful data. Finally, the systems were to include feedback loops so that they not only ensured candidate performance, but also that the performance data were looped back into the system for overall and ongoing program improvement.

Clearly the changes expected by the IPSB were significant for the campuses and for the teacher preparation programs. During this time, Indiana received a Teacher Quality Grant through Title II federal funding that helped support the process; however, relatively little other funding was in place for design and implementation, either at the state or institutional level.

Although funds were minimal, there was strong motivation to participate in the process. First, of course, program approval from IPSB was at stake. Additionally, the work was aligned with the National Council for Accreditation of Teacher Education (NCATE), a voluntary accreditation agency for teacher preparation programs. Working collaboratively with INTASC and the National Board for Professional Teaching Standards (NBPTS), NCATE was phasing in a requirement for institutions to develop and implement "unit assessment systems" for teacher preparation programs, and Indiana was frequently credited with informing NCATE's development of these systems. It is noteworthy that, although the IPSB did not require programs to achieve national accreditation through NCATE, the Board adopted NCATE's standards for program approval. Consequently, nearly all institutions with teacher preparation programs opted for national accreditation through NCATE. The alignment between national accreditation and state program approval processes was important because it limited duplication of reporting requirements for institutions. Additionally, there was a common under-standing about assessment systems, ongoing improvement, and accountability. Interestingly, and almost a precursor to language used in education reform circles today, the focus and rhetoric of these efforts was on what teachers could do in the classroom to impact student learning.

Work was also underway on the "third floor," a performance-based licensing system with parallel standards and increasing levels of performance from entry in the profession through ongoing professional development. By 2000, a two-year teacher induction process had been designed, informed largely by INTASC and a highly regarded induction model from the state of Connecticut. Following the model for input that had been used previously, an advisory group, made up of selected volunteers from across the K–16 spectrum, with IPSB staff support, designed the comprehensive induction system. Their recommendations, in turn, went to the Board for approval. Beginning teachers were to receive support during their first two years of employment in their schools from a trained mentor teacher. Mentors were not to serve in a supervisory capacity, but rather more as a coach and resource. Prior to receiving a license at the end of the second year of teaching, beginning teachers were to create a portfolio with evidence to support their ability to teach effectively. The evidence would include lesson plans, a video, and reflective statements. The focus was to be on effectiveness in impacting student learning. The portfolio, in turn, would be scored by experienced teachers from across the state who had been trained in scoring the portfolios by the IPSB. Novice teachers who were not successful with the high stakes assessment would generally have the option of one additional year in the classroom; however, if they were not successful at that juncture, they would not receive a license to continue teaching. This model was significantly different than the one in place—namely, a form, signed by school district leadership, recommending continued licensure.

It became clear during the pilot phase that the time involved in training scorers and the logistics involved in distributing and collecting the portfolios and recording scores would require significant additional resources from the IPSB. Additionally, concerns were voiced from teachers about the amount of time required for beginning teachers, who were already working long hours, to complete the portfolio process. At the heart of the debate was whether or not the beginning teacher portfolio process served to support professional development, especially reflection on practice, or if it was merely a hoop that new teachers were required to jump.

The next stage was being planned. It was focused on the development of individual growth plans for teachers as part of a performance-based licensing system. These growth plans would span a teacher's career, and be linked to ongoing licensing. Part of the plan was that experienced teachers could gain an "accomplished teacher license" in one of two ways. A teacher could gain certification from the National Board for Professional Teaching Standards or obtain an advanced degree from an institution approved by the Indiana Professional Standards Board. The latter was seen as being viable, given that the institutions were focused on the Indiana standards and were beginning to implement performance-based assessments, at the graduate levels.

These changes were dramatic—they represented a fundamental shift away from licensure being automatic based on submitting paperwork and time in the classroom to licensure based on documented performance in the classroom and ongoing professional development.

The System: An Institutional Perspective

At institutions of higher education in Indiana, the development and implementation of unit specific performance-based assessments began to take hold, but with varying degrees of intensity throughout the late 1990s. A few institutions were satisfied with the status quo and did not share the Board's vision. In other cases, university administrators were concerned about the time and costs associated with the changes. The majority of institutions were engaged and committed to the notions of documenting student performance and institution accountability, albeit to varying degrees, and with varying concerns about associated resources needed for design and implementation.

As was noted in a 2000 NCATE publication, at the University of Indianapolis, where this author served as dean, elements of the system were been being pilot-tested extensively, and each year more elements became a part of the system of expectations for teacher candidates. The program ensured that candidates' skills were assessed at specific points during the preparation program. If candidates' skill levels were not acceptable, they were referred for remediation or counseled out of the program. For example, at the end of the sophomore year, prior to formal admission to the teacher preparation programs, candidates were required to spontaneously compose a written response to a well-crafted case study, and gain a satisfactory rating on an interview with clinical and school faculty strictly controlled by inter-rater reliability measures. The case study was designed to elicit information about a candidate's dispositions and the spontaneous writing was designed to measure ability to write coherently without the opportunity for revisions. The latter was an example of the institution's use of data to inform ongoing improvement. In this case, feedback from supervising teachers indicated that student teachers were not adept at the real world expectation to compose without opportunities for revision. Consequently, based on data, faculty from the English Department (where coursework in composition was housed) were engaged and strategies/expectations modified to ensure that candidates would have the skills necessary to craft the type of written communications (e.g., to parents, administrators, etc.) expected in a typical classroom. In 1999, the case study/interview performance element was formally embedded into the expectations for candidates prior to admission to the program at the University of Indianapolis. In the junior year, elementary methods courses were taught on site and candidates experienced direct application of concepts taught to students in P-12 classrooms. This experience provided another assessment point to determine whether candidates would continue in the program. Admission to student teaching and student teaching itself were also points of assessment. In order to pass student teaching, candidates completed an "INTASC-like" portfolio including analysis of videotaped performances and evaluation of student work samples. Portfolios were scored by university faculty (both in education and arts and sciences) and K-12 teachers. Scorers were provided training and results were analyzed to ensure inter-rater reliability.

As was the case at the state level, there was little funding at the university to support the design and implementation of the individual assessments or the overall design of the unit assessment system. Although this work was honored as part of the promotion and tenure process for faculty and provided significant opportunities for research and publication, most faculty members maintained a full teaching/advising load throughout the design and implementation of the new system. K-12 teachers also volunteered time to the creation and implementation of the system. Teachers were given small gift cards as tokens of appreciation for scoring portfolios, but since each portfolio was estimated to be two to three hours of work, a $25 gift card was, indeed, a token. Although there was widespread agreement about the importance of shared language and high expectations across campus, the additional load on faculty was considerable and over time the value-added began to be questioned. Additionally, the need for data management systems continued to increase. Debates ensued about whether the university should design its own assessment system or purchase one from vendors. The university opted to contract with an outside consultant to create a data system. In the end, however, the in-house system did not have the necessary capacity or reporting ability and an external product was purchased. The process illuminated a need for clarity about roles/responsibilities/decision making between university information technology staff and the teacher education program.

A House Divided

Initially, the Professional Standards Board enjoyed tremendous success and respect within the state. The vision was seen as important, perhaps in part because the Executive Director was an effective communicator who was skilled at working with a range of constituency groups. She demonstrated an awareness of the politics at play, and worked in collaborative fashion with the legislature, governor's office, superintendent's office, school districts, and universities (NCATE, 2000). There can be little doubt that another key ingredient of the initial success stemmed from the care taken by the governors in office during this time and their staffs to appoint persons to the board who were committed to the IPSB's mission of developing high standards and a thorough, comprehensive process for program approval and teacher licensure. Indiana was seen as a leader in establishing systems that would lead to comprehensive reform (NCATE, 2000).

From the outside, the Professional Standards Board appeared to be flourishing; however, concern was mounting in various sectors that IPSB exercised too much power, was becoming overregulatory, and, in the words of some, a bloated bureaucracy. Although she had been part of the process from the beginning, the state superintendent of K-12 sought more control over educator preparation and licensure. By contrast, the IPSB viewed its autonomy from the state department of education as a professional necessity. External voices predicted that the system was going to collapse under its own weight. Specific concerns were voiced about the complexity of the 17 standards, the complexity of the unit assessment systems, the human resources needed to score and record results from the teacher induction portfolios, and whether or not, in the end, the result would be a better teacher workforce. Additionally, there were concerns about costs. Resources had not been allocated at either the state or institutional levels to support the demands of a totally new system. Faculty members from both K-12 and higher education were asked to serve in a voluntary capacity on numerous committees and task forces This was positive in terms of inclusion; however, as noted in the University of Indianapolis example, those same people were being asked to implement massive change at the institutional and K-12 levels with little, if any, support.

In 2000, the Executive Director resigned for personal reasons. In a significant development, the governor did not seek a permanent replacement for over a year. Although an internal staff member was appointed in an acting capacity, the Board, which was at a very pivotal point in launching massive reform, was essentially without a leader. Additionally, key appointments to the board itself were left open. The question remains open about whether this was a strategy to weaken the board or simply not seen as a priority for the governor.

When a new director was finally appointed in 2001, core work had been maintained; however, due to extended lack of leadership, limited resources, and the reality that attention of key leaders in higher education and K-12 was waning, the reform agenda had begun to weaken. Many of those across the K-16 spectrum who had played key leadership roles over the last few years in the reform effort either took new positions or assumed new responsibilities during the lapse in leadership, creating a void in the capacity that had been skillfully crafted by the prior Executive Director.

In 2005, control of the governor's office shifted from one party to the other. The state superintendent of K–12 schools, who had been in office since the inception of the IPSB, was elected to a fourth term. Indiana now had a governor and state superintendent from the same party, leading some to believe the stars were aligned. It quickly became apparent that their differences about education were significant. One thing that they agreed on, however, was that change needed to occur with the IPSB. Whether in a move to regain control on the part of the superintendent, a move to eliminate a bureaucracy on the part of the governor, or a variation on both, in July 2005 the Indiana Professional Standards Board was abolished and the functions were once again placed under the auspices of the Department of Education.

Although in a position to change the direction that had been forged by IPSB, the state superintendent did not push additional development of the existing reform agenda, nor did she make major changes to the framework or standards adopted by the IPSB. A 19-member advisory board, with the same composition as the original IPSB, was maintained; however, delay with new appointments to the board by the governor resulted in minimal board action. Consequently, although the Board was no longer autonomous and was now under the control of the Department of Education, the overall approval processes for teacher preparation and the licensure changes remained in effect. Few, if any, additional steps were taken toward completing the overall vision. For example, no changes were made to how teachers received support in their first two years in the classroom, nor was the intended portfolio process of evaluation for continued licensure implemented.

After 16 years in office, and, as the *Fort Wayne Journal Gazette* noted, under pressure from the governor, the state superintendent did not run for re-election in 2008 (*Journal Gazette*, 2008). After running on a platform demanding accountability with focus on classrooms and student achievement, a superintendent of the same party affiliation as the governor was elected and aggressively took charge in 2009. By most accounts, the new superintendent shared the governor's belief that less is more.

As an example, in a 2009 interview with WNDU, South Bend, Indiana, the new superintendent justified a reform strategy of changing licensing requirements. The news station explicated the changing requirements would mean that "an education degree will no longer be enough for Indiana teachers" (McFadden, 2009, para. 6). The superintendent elucidated: "They were not subject to that rigorous content knowledge in college ... New licensing regulations would provide schools the flexibility so that they have teachers with the content knowledge to move around, to put with the children they need to put them with in order to get their needs met" (McFadden, 2009, para. 7). The Indiana superintendent further clarified:

> We're not going to abandon those very important "how to teach" courses, but we are going to say there should be a more strenuous focus on content ... Our children are becoming incredibly competitive nationally and internationally, and in order to be competitive you have to have a rich content knowledge.
>
> (McFadden, 2009, para. 9)

He concluded his remarks by saying teachers should receive:

> The training and recertification they need through their own school districts through workshops and classes ... Putting the onus on the school corporation, the building level administrators, who really are responsible for teacher quality ... Our intent is to every day find ways to take Indiana from where it is to the place where every business says "we have to move to Indiana."
>
> (McFadden, 2009, para. 14)

The changes were dramatic. According to the Indiana Department of Education website, no meetings of the Professional Standards Advisory Board were held for the first five months of 2009. The Board met in June 2009, with an agenda that focused on a presentation from the National Council on Teacher Quality "regarding national perspectives and how Indiana policies may impact teacher quality" (Professional Standards Advisory Board, 2009, p. 2). The Board met monthly through the end of 2009; however, according to the department's website, only one meeting was held during the first six months of 2010. The new rules for teacher licensing and preparation were designed during the last six months of 2009. Labeled as Rules for Educator Preparation and Accountability (REPA), they were approved by the Board in January 2010, and signed into law by the governor in March of the same year. According to the Indiana Department of Education website, the new rules were designed to "take steps to address future teacher shortages and bring more knowledgeable adults into Indiana schools" (IDOE, 2010, para. 6). The website noted that:

> The advisory board will have the authority to approve online and non-traditional preparation programs in the future. Without these alternative licensing programs, it is unduly difficult for successful adults in other careers to enter the teaching profession. The new regulations allow for new pipelines to bring real world experts into Indiana classrooms.
>
> (IDOE, 2010, para. 6)

The new regulations were to allow for teachers to add subjects to their licenses solely by passing content knowledge exams, in stark contrast to the performance-based systems designed by the IPSB.

What were the effects on the third floor and the status of aligned licensing including the teacher induction model? The Indiana Department of Education website stated:

> After receiving an Initial Practitioner License beginning teachers, administrators, and school service personnel will participate in a two-year period of mentorship. The culmination of the mentorship is the completion of a specific assessment piece.
>
> <u>Year Two Teachers and School Service Personnel</u> will complete the assessment piece with their Building Level Administrator, Principal, Supervisor, or Director. **A portfolio is no longer required.**
>
> <u>Year Two Administrators</u> will complete the online self-assessment through Moodle and submit the Individual Development Plan Summary Form. Once these requirements have been met, Year Two IMAP candidates will be eligible for the five year Proficient Practitioner license.
>
> (IDOE, 2009, paras. 1, 2, 3; underlines and bold as in original)

The process was consistent with the changes made in teacher preparation, namely streamlining of processes with an emphasis on getting teachers into classrooms with the greatest expediency possible.

The changes were contested by teachers and university personnel. One College of Education administrator said REPA "has the effect of deregulating and de-professionalizing our business," a sentiment that was echoed by a school corporation official, who explained, "the proposed changes 'are devaluing the teaching profession'" (Loughlin, 2009, para. 15; Loughlin, 2009, "Teaching is more" section, para. 3). In voicing concerns about the new rules, the president of Indiana State University requested the Department of Education to work with the educational community to bring about change. Although one superintendent expressed that he felt "the REPA discussions demonstrate[d] that officials listened to testimony from various professional groups and individuals, to some extent," others fervently disagreed (McCollum, 2009, para. 4). Despite public hearings on the proposed changes, "some teachers [said] the time of the hearings at 10am makes it difficult for them to attend because school is in session," and one educator went as far as to say, "the state 'is slipping things through without letting people really know. We're not informing the people this really impacts'" (Loughlin, 2009, "A different point of view" section, para. 3; Loughlin, 2009, "Teaching is more" section, para. 2). The Indiana Association of Colleges for Teacher Education, in a memorandum dated August 26, 2009 to the Professional Standards Advisory Board (IACTE, 2009), noted five concerns, including restrictions on professional education credit hours, the omission of academic and professional standards, a program approval process without standards and accountability, and an unrealistic timeline (p. 1). The concern for the reduction of pedagogical courses emanated from a variety of stakeholders, especially those from the school corporations. The

Indiana Association for Health, Physical Education, Recreation and Dance issued a letter stating, "licensing teachers based heavily on content-knowledge seems to put the focus on the information and not the students;" the organization went on to give the example of how someone could pass a written examination on physical education and still not be trained to go through activities with children to ensure minimal injury (Loughlin, 2009, "Teaching is more" section, para. 14). One school corporation superintendent explained, "there must be some assurance of crossovers from other professions having some sense of teaching methods" (McCollum, 2009, para. 7), and one teacher articulated the matter quite bluntly:

> Being an expert in a certain area doesn't necessarily make that person a good teacher ... Expertise must be accompanied by a knowledge of teaching methods and what works with different children, who have different learning styles. Those who want to be teachers also need to learn classroom management.
>
> (Loughlin, 2009, "Teaching is more" section, para. 4)

Although there is little evidence of an inclusionary model being utilized in the development of REPA, the Indiana Department of Education website noted that the superintendent planned to host a series of statewide meetings in August 2010, to build understanding.

The Future

The ultimate question, and one for which little, if any data exists, is the effect of the elimination of the autonomous IPSB, and its reform agenda, on teacher quality in Indiana. Without a doubt, there has been a dramatic shift from professional standards and complex assessments to employment expediency.

The effect of such these moves underscores the need for the development and implementation of a fully integrated longitudinal data system. Such a system, if well designed and implemented, will provide policymakers, school districts, and universities much-needed data about effective practices in teacher preparation, induction, and professional development—practices that result in increased student achievement. Preliminary data from North Carolina indicates that there is as much variance within program types (traditional versus "non-traditional") as between them (Henry, Thompson, Fortner, Zulli, & Kershaw, 2010). Such evaluation and research is critically important in states such as Indiana, which have taken a relatively unregulated stance toward teacher preparation.

Changes, of course, have occurred at the national level as well. As an example, in July 2010, the Council of Chief State School Officer's (CCSSO) Interstate Teacher Assessment and Support Consortium (acronym changed from INTASC to InTASC) released "Model Core Teaching Standards: A Resource for State Dialogue" for public comment. The document represented a major revision to the standards for beginning teacher assessment and development that were heavily used by Indiana in the early 1990s. No longer limited to the assessment and support of new teachers, the 2010 standards articulate standards of professional practice for all teachers.

In the meantime, university-based teacher preparation programs in particular need the freedom to innovate, learn, and compete with new alternative programs. Policies must demand accountability of any program provider, whether university based or alternative. A major question remains as to how regulatory entities, whether through Departments of Education or autonomous Professional Standards Commissions, can meet their accountability responsibilities and simultaneously create a climate that is innovation friendly and result in demonstrated impact on teacher effectiveness.

References

Darling-Hammond, L. (1997). *Doing what matters most: Investing in quality teaching.* New York, NY: National Commission on Teaching & America's Future.

Henry, G.T., Thompson, C.L., Fortner, C.K., Zulli, R.A., & Kershaw, D.C. (2010). *The impact of teacher preparation on student learning in North Carolina public schools.* Chapel-Hill: Carolina Institute for Public Policy.

Indiana Association of Colleges for Teacher Education (IACTE). (2009, August 26). Letter to Professional Standards Advisory Board members (Word document). Retrieved from www.iacte.net/node/58

Indiana Department of Education. (n.d.). Standards. Retrieved from www.doe.in.gov/educatorlicensing/preface.html

Indiana Department of Education. (2009). Indiana mentoring and assessment program (IMAP). Retrieved from www.doe.in.gov/educatorlicensing/IMAP.html

Indiana Department of Education. (2010). New rules will give Hoosier students more knowledgeable teachers: Subject-experts from outside education welcome. Retrieved from http://www.doe.in.gov/news/2010/01-January/NewRulesKnowledgeableTeachers.html

Journal Gazette. (2008, February 20). Reed should seek re-election (Editorial). Retrieved from www.journalgazette.net/apps/pbcs.dll/article?AID=/20080220/EDIT07/802200371/-1/EDIT

Loughlin, S. (2009, October 31). Debate heats up over revamped teacher license rules. *Tribune Star.* Retrieved from http://tribstar.com/local/x546422483/Debate-heats-up-over-revamped-teacher-license-rules

McCollum, C. (2009, December 11). State makes chances in license rules for teachers. *Times of Northwest Indiana.* Retrieved from http://www.indianaeconomicdigest.net/main.asp?SectionID=31#x0026;SubSectionID=135&ArticleID=51440

McFadden, M. (2009, August 19). A conversation with Indiana superintendent Dr. Tony Bennett, part 1. *WNDU.* Retrieved from www.wndu.com/specialfeatures/headlines/53613487.html

Professional Standards Advisory Board. (2009). *Minutes* (PDF document). Retrieved from www.doe.in.give/educatorlicensing/pdf/PSBMinutes6-25-09.pdf

Scannell, M. & Wain, J. (1996). New models for state licensing of professional educators. *Phi Delta Kappan 78,* 211–215.

EDITORS' COMMENTARY

Establishing standards boards in teaching can be seen as a struggle between those wanting to take a step toward making teaching a profession and those who believe that the decision as to who will teach belongs squarely within state government—in this chapter, the history of the Indiana Standards Board, which could be viewed as an example of the rise and fall of an independent standards board is presented.

The National Education Association has asserted that it is important for a profession to control itself. The organization has called for the establishment of independent professional standards boards in every state and that the boards be authorized to create standards for certifying teachers and issue certificates to them. The NEA continues its stance, arguing that a profession must control itself. The organization calls for the establishment of independent professional standards boards in every state, which are authorized to create the standards for certifying teachers and issue certificates to them.

The question of who controls the field, or profession, is perhaps the only quality assigned to a profession that can be achieved by a simple change of the laws. We note that there are other qualities usually attributed to a profession, including (Shulman, 1987/2004):

- the obligation of service to others, as in a calling;
- understanding of a scholarly or theoretical kind;
- a domain of skilled performance or practice;
- the exercise of judgment under condition of unavoidable uncertainty;
- the need for learning from experience as theory and practice interact;
- a professional community to monitor quality and aggregate knowledge.

If these are the qualities of a profession, one could argue whether or not teaching qualifies as a profession, and this is indeed an ongoing argument. Some would argue that not only is teaching not a profession, but it should not become a profession. Can we assume that when states establish independent standards boards, as Indiana did, that the state recognizes teaching as a profession?

Here we have a case of one of the earlier independent standards boards, but not the first. The first was the California Commission on Teacher Credentialing, created in 1970 by the Ryan Act. It was formed to address the myriad of problems having to do with an insufficient number of teachers for a growing school population. Intended to help professionalize teaching, the standards board was established independent of the California Department of Education as an autonomous board to set high standards for the initial and continuing education of teachers. It was established with the belief that a self-governing teaching profession would enhance the quality of education in California. Teachers were to have a prominent "say" in making teaching more respected and better compensated. The successive 30 years are replete with efforts to broaden the definition of professionalism and to find other vehicles to professionalize teaching.

Of course, we know that the Indiana Standards Board, once lauded for its independence and leadership, no longer exists. What is the status of standards boards nationally? Those remaining include California, Delaware, Georgia, Hawaii, Iowa, Kentucky, Minnesota, North Dakota, Oregon, and Wyoming. Semi-independent boards are found in Nevada, North Carolina, Delaware, Maryland and Texas.[1] Interestingly, the updated identification of standards boards in the United States comes from a report that studied the state auditors recommendation to eliminate the Hawaii independent board.

One thing the case clearly illustrates is the important role strong leadership played in the Indiana Board. The conception of this kind of board, however, ultimately depended on the support of the governor and state superintendent (elected in Indiana). When both officials changed, support for the board disappeared.

A clear paradox plays out in Indiana as a board born of one set of premises about teaching and professionalism saw support for that view evaporate, perhaps nationally as well as in the state. How do we make sense of such a dramatic move in agenda? We cannot, of course, know all the motivations of the actors, nor can we necessarily ascribe a position to one political party or another. Some argue that the Race to the Top program, embodying policies of a Democratic adminis-tration, is anti-professionalism because it narrows the measure of successful teachers, does not allow the "profession" to determine success, and that it advo-cates non-university sites for teacher education.

Discussion Questions

1. Do you think teaching meets the criteria to be called a profession? Is inde-pendent control of entry into the profession by teachers critical to its standing as a profession?

2. Besides perhaps helping the evolution of teaching as a profession, are there other possible motivations for creating an independent board beyond a goal of professionalism? Who would be motivated by your list? Are there any budgetary considerations?

3. Are we engaging in wishful thinking if we advocate independent boards? Can you think of other cases where state government, or federal government, has voluntarily given up control and power?

4. Do you think that having a teachers' union advocate for professionalization helps or hurts the probability of teaching being recognized as a profession? Is this a good tool to change policy?

5. The author wrote that reform in Indiana had "shallow roots." Is there any example in the other chapters where reform had deep roots? Why is the idea of institutionalizing teacher education reform so illusive?

6. The induction and mentoring plan created in Indiana in some ways parallels the teacher residency play supported by the Obama administration. Does the Indiana experience inform the challenges associated with implementing residency programs?
7. It seems that the reform agenda in Indiana was never truly institutionalized. Is this apparent in other chapters?

Note

1 In the NEA's classification, there are a series of conditions that must be met for a board to be independent, ranging from authority over standards, budget, fees, the authority to issue, renew, and revoke licenses, and the authority to approve teacher education programs. Those listed as semi-independent lacked one or more of these elements of authority.

7

PARTNERSHIP FOR TEACHER EDUCATION

The Case of Montclair State University and its School-University Partnership

Ada Beth Cutler, Frank Alvarez, and Susan Taylor

The story of school-university partnership at Montclair State from 1986 to 2010 is rich with purpose, serendipity, and hard work that enabled these partnerships to grow and flourish, aided and abetted by the ebb and flow of federal, state, and public policies aimed at shaping the nature and role of partnerships for teacher education. We tell this story from three perspectives: the university, an urban school district partner, and a suburban school district partner. Collectively, as faculty member and director of the New Jersey Network for Educational Renewal (the partnership organization) and then dean of the College of Education and Human Services, principal and then superintendent of Montclair Public Schools, and teacher and then principal in Newark Public Schools, we have over 50 years of experience with partnership between Montclair State and its nearby public schools.

To tell our story, we first lay out some of the history and milestone policies, events, and programmatic developments in the partnership from the University's perspective and then we focus on how this history played out in two member districts in particular. Viewed through the lens of almost 25 years of formal partnership, our story is one of genuine progress, significant accomplishments, and positive momentum for the schools, for teacher education, and, most important of all, for the children in our partner schools. And yet, throughout those years, there were moments of disappointment, conflict, and backsliding and we hope to portray these as well as the overall achievements, in the interest of honesty, education, and accuracy.

Partnership from the Perspective of Montclair State University

Formal Partnership Begins

The year 1986 carried major milestones for school-university partnership, both locally and nationally. The Holmes Group, a consortium of major research universities engaged in teacher preparation, released its first and arguably most seminal report, *Tomorrow's Teachers* (1986), aimed at reforming teacher education and the teaching profession. The Holmes Group formed as a reaction to the widespread, national discontent with public education and teacher preparation fueled by the alarmist report, *A Nation at Risk*, published in 1983 (National Commission on Excellence in Education, 1983). Central to the Holmes Group's vision for reform was the call for professional development schools (PDS)—akin to teaching hospitals in medical education—that would be exemplary schools in formal partnerships with higher education institutions, where university faculty and public school teachers and administrators would share responsibility for teacher preparation, inquiry, innovative practices, and the professionalization of teaching. The notion of public schools as central players and partners in teacher education, rather than merely sites that accept and host student teachers, was a radical departure from past practice in teacher preparation, especially among elite research universities.

The 1980s were also watershed years for teacher preparation at Montclair State University, in New Jersey, called Montclair State College (MSC) at that time. Founded in 1908 as the second normal school in the state, Montclair had a long history of excellence in teacher education, but the policy climate in New Jersey in the 1980s was marked by disrespect and even disdain for university-based teacher education (Klagholz, 2000). The first alternate route to teaching was introduced in New Jersey and regulations were changed to create one blanket certification for Grades N-8. Despite the difficult times for teacher education in the state, in 1986 Nicholas Michelli, education dean at Montclair State since 1980 and a faculty member for ten years before that, made the decision with faculty support to designate Teaching for Critical Thinking the theme of the teacher education program, building on the strengths of its faculty and programs such as the Institute for Critical Thinking, the Institute for the Advancement of Philosophy for Children, and Project THISTLE (Thinking Skills in Teaching and Learning) in the Newark Public Schools which had begun six years earlier. To ensure congruence between campus courses and clinical experiences in the schools around Teaching for Critical Thinking, 1986 also saw the launch of the Clinical Schools Network, a fledgling school-university partnership dedicated to providing professional development for partner schools on Teaching for Critical Thinking. In many ways, these two advances at Montclair State laid the foundation for future innovations, grants, programs, and networks that contributed to Montclair's preeminence in school-university partnership and collaborative teacher education across the teacher development continuum.

The Clinical Schools Network consisted of five nearby school districts that accepted large numbers of MSC student teachers and the College. In addition to providing no-cost professional development, the dean enabled the Network to appoint qualified cooperating teachers from partner schools as Clinical Faculty members of the University, with campus ID, library access, and faculty parking tags as privileges. Newark, the largest school district in the state with high concentrations of poverty and African-American and Latino students, and Montclair, a suburban school district with longstanding racial diversity, were among the original members of the Clinical Schools Network and for 24 years they have been important partners in teacher education and school renewal with Montclair State.

The Harold Wilson Professional Development School

In 1990, with the growing currency and cachet associated with professional development schools, Montclair State entered into an intensive partnership with the Newark Board of Education and the Newark Teachers' Union to establish the Harold Wilson Middle School for Professional Development. Based on the Schenley School model from Pittsburgh, this PDS served as a professional development hub for middle grades teachers in the district, with a cadre of replacement teachers who took over teachers' classrooms while they were in residence at Harold Wilson. Although the PDS was dismantled one year after the state of New Jersey took control of the Newark Public Schools in 1996, there were lasting effects of this partnership at Montclair State and in the Newark Public Schools. Many faculty members, including the current dean Ada Beth Cutler and the current director of the Center of Pedagogy at Montclair State, Jennifer Robinson, cut their professional development teeth at Harold Wilson and formed lasting relationships with educators in the district who now play leadership roles and help sustain the partnership between MSU and NPS.

Montclair State's Membership in the National Network for Educational Renewal

In 1991, building on his groundbreaking 1990 book *Teachers for our Nation's Schools,* John Goodlad solicited applications for the newly reconstituted National Network for Educational Renewal (NNER), and he made partnerships with local districts a requirement for all applicants. Dedicated to the simultaneous renewal of the schools and teacher education and based on the Agenda for Education in a Democracy, the NNER provided a moral and philosophical framework for partnerships between schools and teacher-preparing colleges and universities as opposed to the reform and professionalization orientation of the Holmes Group. The Agenda for Education in a Democracy laid out a four-part mission of public education and teacher education. These are: fostering in students the habits and skills necessary for participation in a democratic society, ensuring that students have access to the knowledge and skills that will enable them to lead full and productive lives, providing teachers who can nurture students' development and learning, and developing educators' commitment to stewardship of best practice. This framework resonated at Montclair State, with its focus on Teaching for Critical Thinking, the Clinical Schools Network, and a commitment to urban education manifest in partnership work with Newark. Dean Michelli and the director of teacher education at Montclair State, Robert A. Pines, saw an opportunity to marry these commitments and programmatic features with Goodlad's Agenda for Education in a Democracy and they applied to become an inaugural member in the NNER. As it turned out, the critical thinking connection was not difficult, given the focus of the NNER on education for democratic participation. Out of hundreds of applications, eight settings were accepted to form the NNER and Montclair State was one of the eight.

It is difficult to overstate the impact of membership in the NNER for Montclair State and its partner schools. Philosophically, the NNER provided a clear and compelling framework for the purposes and required elements of teacher education and public education in a democratic society. John Goodlad posited that there are three groups responsible for teacher education and dubbed these three—education faculty, arts and sciences faculty, and school faculty—the tripartite. Furthermore, he proposed the concept of simultaneous renewal of the schools and teacher education as the focus of the NNER and laid out the Agenda for Education in a Democracy (AED) as the ethical and philosophical grounding for the work of simultaneous renewal. At Montclair State, members of the tripartite gathered frequently to unpack, argue, and develop their own understanding of the Agenda and Goodlad's 19 postulates that defined the necessary conditions for high quality teacher education (Goodlad, 1994). Over time, a focus on social justice and urban education at Montclair State added meaning and local nuance to the Agenda and fostered ownership of the Agenda on campus and in partner schools. Even now, almost 20 years later, the Agenda for Education in a Democracy is writ large in policy and practice within the College of Education and Human Services at Montclair State and its school-university partnership.

Politically and practically, membership in the NNER was seminal for the ongoing development of teacher education and the concept of partnership with local schools at Montclair State. First, John Goodlad's national reputation as one of the 20th century's foremost educational researchers, change agents and theorists brought new visibility and prestige to teacher education both internally at Montclair State and externally within the larger world of schools of education. Internally, the dean of education brought the Provost, Richard Lynde, to the University of Washington to meet John Goodlad, whose charisma and deep intellect impressed the Provost. The Provost, who served in his post from 1987 to 2008, became a champion for the importance of the NNER at Montclair State. That translated to a greater willingness on the part of the Provost and President to provide funding for various elements of teacher education required by the NNER, such as the school-university partnership. Even in difficult budget years, funding for teacher education not only stayed steady but often increased with the development of new structures and initiatives, such as the Center of Pedagogy and the Teacher Education Advocacy Center. In essence, through membership in the NNER, Montclair State was garnering national attention for its teacher education program and the President and Provost recognized and appreciated this and thus chose to support teacher education with robust funding for innovation and excellence.

In 1994, Montclair State College became Montclair State University, signaling its growing stature as a comprehensive institution of higher education and opening the door for doctoral programs in the future. An historical note of interest is that the first doctoral program at the University, an Ed.D. in Pedagogy with a set of core courses based on the Agenda for Education in a Democracy, was launched in 1998. With the advent of university status, the School of Professional Studies that housed teacher education became the College of Education and Human Services in 1994. That same year another important name change occurred. The Executive Committee of the Clinical Schools Network at Montclair State voted to change the Network's name to the New Jersey Network for Educational Renewal (NJNER) in recognition of its growth from five member districts to twelve. It later became the MSU Network for Educational Renewal in 1999 to alleviate confusion in the state and to focus on the University as the Network's hub. Also in 1994, its first full-time director, Ada Beth Cutler, came to Montclair State as a faculty member in Curriculum and Teaching. Initially, University funding for the NJNER's activities came through the Institute for Critical Thinking, but with the hiring of a faculty member as the first director of the partnership, the dean established a regular budget account of $25,000 for materials, mailings, and professional development activities.

A multi-million-dollar grant from DeWitt Wallace in 1994–1997, funneled from the NNER to member settings, provided a major shot in the arm for the NJNER. The grant funded a Graduate Assistant, Teacher Study Groups, Summer Institutes, and a series of mini-courses including two on Teaching for Critical Thinking and Mentoring and Coaching required for all Clinical Faculty members. The DeWitt Wallace grant required a $75,000 match each year from the University, which the provost granted and eventually made a permanent line in the budget (Patterson, Michelli, & Pacheco, 1999). This expansion of the professional development programs within the NJNER enabled the partnership to begin to charge nominal membership dues for partner districts ($2,000–3,000 a year depending on the size of the district), providing a new infusion of money for the Network. But more than the infusion of money the move to charge dues demonstrated shared responsibility between the university and the school districts and, perhaps, something for which a fee was charged was seen as more valuable than a "free" service. Additionally, the NJNER garnered its own grants for professional development programs from the Geraldine R. Dodge Foundation and other local foundations, adding almost $1 million to the coffers over ten years. Each year, the dean was able to add incrementally to the university's regular budget for the Network. The annual budget from the university and membership dues grew from $25,000 to over $200,000 in 15 years. By 2009–2010, 27 member districts were contributing over $100,000 in membership dues a year. All of the programs originally funded by the DeWitt Wallace grant thus became permanent features of the Network, and many new initiatives were added over the years, including online professional development offerings. Literally thousands of teachers in member districts take part in Network professional development programs each year at no cost to their school districts beyond membership dues.

The partnership grew not only its resources and funding base, but also its culture and structures. Each district appointed one or two District Coordinators, who served as the formal liaisons to the university for the Partnership and formed the NJNER Operations Committee, which met monthly to plan activities and formulate policies to be voted on by the Executive Committee. The MSUNER began to pay these District Coordinators annual stipends, with some receiving additional stipends from their districts to play a formal role in field experience placements. The Operations Committee established a rule that all professional development programs or activities had to be jointly offered by MSU faculty and school faculty, in recognition of the fact that not all of the wisdom, knowledge, and skills for staff development resided at the university. Up until that point, only university faculty delivered professional development, creating an element of hierarchy the partnership deliberately eschewed. At the same time, parity of pay was established and facilitators of professional development programs were compensated at the same rate regardless of degrees or home institutions.

In 1995, the first Center of Pedagogy in the nation was established at Montclair State, to oversee initial teacher preparation at the university; bring together education, arts and sciences and school faculty and administrators on neutral ground, as the "tripartite" group responsible for teacher education; and to provide an appropriate home for the NJNER at Montclair State. By 1996, the Network had grown to include 20 school districts, with a variety of sizes and demographic profiles.

The Age of Professional Development Schools

By the early 1990s, the concept of partnerships for teacher education, especially in the form of professional development schools, had captured the attention and imagination of policymakers and funders at the national and local levels. Policymakers at the national and state level were pushing for partnership as a central feature of teacher education and the pressure to declare schools PDSs was real and compelling. This also led to an infusion of funding opportunities for PDSs from agencies and organizations such as the New Jersey Department of Education, the National Education Association, and the Holmes Partnership (formerly the Holmes Group). At first, Montclair State embraced these opportunities to establish new PDSs. The visibility of the Harold Wilson PDS and the larger NJNER partnership helped Montclair State garner its share of this funding. A PDS grant from the New Jersey Department of Education, funds and recognition from the NEA-Teacher Education Initiative (Montclair State was one of nine settings in the project), and membership in the Holmes Partnership led to an explosion of PDS partnerships within the NJNER.

Three high schools, one middle school, and three elementary schools established formal PDS agreements with MSU (including Montclair High School) during the 1990s. Exciting developments ensued at these PDSs—with new opportunities for growth and learning and research for MSU and PDS students and faculty. One school instituted "rounds" long before it gained cachet; another created social justice inquiry projects for students guided by MSU student teachers and faculty; and another established a Civics and Government Institute with the help of MSU faculty from the arts and sciences. The halo effect was alive and well and, despite the enormous commitment of time and funds, the PDSs flourished initially.

An overall NJNER PDS Council, PDS Governing Councils at each school with tripartite representation, the commitment of MSU faculty to serve as PDS Liaisons at each school, the expectation of courses taught on site including graduate courses for teachers, and other features of PDSs combined with the end of grant funds from the state and NEA-TEI, brought complexity, bureaucracy, high expectations, and expense to the PDS endeavor within the NJNER. As inevitable leadership changes occurred at the PDS schools and in some superintendents' offices, and University faculty members' energy and commitments waxed and waned, the bloom was off the rose at most of the PDSs. PDS principals expected an MSU faculty member to be on site as a PDS Liaison and the University expected significant numbers of student teachers to be accepted at PDSs and there were instances when one or the other expectation was not being met. Within some of the PDS partnerships, angry exchanges and bitter feelings ensued.

To make matters worse, jealousy and rivalries developed between and among PDSs. For the most part, these tensions were the result of perceptions of differential resources and commitment on the part of the University and to public recognition won by one or more of the PDSs. Furthermore, "non-PDS" schools within the NJNER that Montclair State depended on heavily for field placements for teacher candidates expressed disappointment and even resentment that they were part of the larger partnership but not privileged PDSs. Within the University, there were resentments between departments and individual faculty members due to differential faculty assignments and course loads at PDSs. In 2001, the new dean who succeeded Dean Michelli and the director of the Center of Pedagogy joked privately about publicly renouncing PDSs and many at the University longed for the days when all the schools they worked with were simply called partner schools, with naturally varying levels of participation in the NJNER and a more organic development of true partnership.

At MSU, faculty members and administrators laughed derisively when colleagues at other universities touted their 25 or 30 or 40 PDSs. It did not seem possible to have so many "real" PDSs, with their resource appetite and requirements, but PDSs had become the coin of the realm, and teacher education programs without them were seen as behind the times and out of step with current expectations. The pressure to declare schools PDSs had become unproductive. At Montclair State, the Network called a moratorium on new PDSs and tried to focus more energy on partner schools that were PDS-like without the formal agreements and requirements. Over time, the preeminence of the PDSs within MSU's partnership faded, some formally shed that appellation, and others continued to function, but without most of the formal structures and requirements. Currently, at MSU, discourse centers only on partner schools and, frankly, most educators involved in partnership from the University and the schools believe all are better off for the change.

Interestingly, the focus on PDSs nationally and at the state level has waned as well. For instance, in 2004, the New Jersey Department of Education held a grant competition for school-university partnerships that did not require professional development schools, but rather focused on developing partnerships that had clear goals of improved teacher education and professional development. Recent federal grant opportunities such as the Teacher Quality Enhancement and Teacher Quality Partnership programs have no mention of PDSs. Clearly there is recognition that partnerships for teacher education do not have to be in the form of PDSs. The MSUNER began before the age of PDSs and it has survived and flourished after their preeminence.

The Ongoing Evolution of the MSUNER and Partnership Endeavors at Montclair State

After the faculty member who directed the partnership became dean in 2000, the Executive Committee of the partnership decided to hire a full-time director of the MSUNER from the schools. A longtime, very active Clinical Faculty member from one of the high school PDSs retired from the schools and became director of the partnership, serving for five years before she retired. She was succeeded by another former Clinical Faculty member from a different high school PDS. Both brought new ideas, energy, and worthwhile change to the partnership, including more professional development offered in the schools rather than on campus, new grant opportunities for Clinical Faculty, a jam-packed newsletter, more emphasis on technology, better preparation for mentoring and coaching, a new required mini-course for Clinical Faculty around culturally responsive teaching, and new partnerships with the Newark Museum and other local agencies. A formal admissions process was developed for new member districts to assure commitment and fit, and as usual the tripartite was represented on the committees that review applications for MSUNER membership. Since 1994, hundreds of funded teacher study groups have spent thousands of hours engaging in self-directed learning; over 50 teams of educators have taken part in the Geraldine R. Dodge Foundation-funded Action Research Groups; over 1,000 teachers have attained status as Clinical Faculty members; seminars and other teacher education courses have routinely taken place in partner schools; thousands of teachers have participated in mini-courses, summer conferences, and workshops; MSU students have volunteered tens of thousands of hours of community service in partner schools; thousands of student teachers have completed legacy projects in the schools they served in, with lasting results such as an annual clothing drive; and two MSU partner schools have won national awards for their achievements as partner schools.

In 2000, the MSUNER provided field experiences to approximately 300 students in their final year of teacher preparation; in 2010, it provided field experiences for over 700 students in their culminating year. Over the years, the professional preparation program for teachers at Montclair State has been transformed and redesigned and renewed with the input and participation of the tripartite group. Clinical Faculty from the partner schools teach courses, mentor students in the field, serve on Center of Pedagogy task forces, conduct research with MSU faculty, and complete graduate degrees at the university. It has been simultaneous renewal in action.

The MSUNER now has 27 partner school districts; a budget from the university of over $100,000 combined with over $100,000 of membership dues from the districts each year; and a revenue-generating arm called MSU-BLCS (MSU Building Learning Centered Schools) that provides intensive consultation, coaching, and professional development to schools in need of improvement. In 2007, *Edutopia*, the magazine of the George Lucas Educational Foundation, named Montclair State one of ten leading schools of education for teacher preparation in the nation, largely due to the strength and depth of its school-university partnership. In 2009, the U.S. Department of Education established a new grant competition, called the Teacher Quality Partnership Program, with $143 million of funding. The rich context of partnership between MSU and the Newark Public Schools was central to the success of Montclair State's Teacher Quality Partnership grant proposal, winning $6.3 million in federal funding along with additional matching funds to establish the Newark-Montclair Urban Teacher Residency Program. Often touted as the new gold standard in teacher preparation, urban teacher residency programs enable post-baccalaureate teacher candidates to spend a full paid year in residence in an urban school while earning a master's degree and initial teacher certification. Over the five years of the grant, Montclair State and the Newark Public Schools will recruit, prepare, and mentor 100 new mathematics, science, and elementary special education teachers for the district. The grant also provides extensive funding for professional development for experienced teachers, adding to the array of opportunities already provided by the MSUNER. Most recently, in 2010, U.S. News and World Report's Best Graduate Schools listed Montclair State's teacher education programs in the top 20 in the nation, based on rankings by deans from doctoral granting institutions. The evolution and development of partnership for teacher education at Montclair State has reaped tangible recognition from the external environment, adding to the University's reputation and resources.

The Partnership from the Perspective of the Newark Public School—Susan Taylor

The Setting

The city of Newark is the largest urban center in the state of New Jersey. The Newark Public Schools (NPS) comprise 80 schools that serve approximately 40,000 students. Most of Newark's students are considered "at risk" because they are from households that fall below the poverty line and qualify for free or reduced-price lunch. Like most large city settings across the country, the Newark Public Schools face multifaceted issues and challenges of urban education related to poverty, such as high dropout rates, students performing below grade level, transiency, unstable home lives, and significant numbers of second language learners.

Currently there are two major overarching policy issues that directly impact and pervade every aspect of the NPS. Both policy issues are based on legal and political decisions that have rippled down from the New Jersey State Capitol building in Trenton and the highest bench of the New Jersey Supreme Court to every NPS central office desk, every classroom and every community within the city of Newark. The implications of both policies directly impact leadership and staffing appointments, fiscal decisions and allocations, academic and curricular choices, and facility design, repair, and maintenance.

The first policy issue is a series of New Jersey Supreme Court decisions in the case of *Abbott* v. *Burke*. This case was filed in 1981 by the Education Law Center in Newark on behalf of an urban 3rd Grade student in New Jersey. In 1985 the first *Abbott* ruling identified 31 urban districts throughout the state as "Abbott Districts." The court specified that Abbott Districts were to receive remedies to achieve what is guaranteed by New Jersey's state constitution—a "thorough and efficient" education for all. The ruling further asserted that public primary and secondary education in poor communities throughout the state was constitutionally substandard. The court mandated that state funding for these districts be equal to per-pupil spending in the wealthiest school districts in New Jersey.

In the Abbott II ruling of 1998 (and in subsequent Abbott rulings), the court ordered the New Jersey Department of Education (NJDOE) to assure that all students in Abbott Districts receive a thorough and efficient education through the implementation of certain reforms, including standards-based education, whole school reform, and full-day preschool for all 4-year-olds, supported by parity funding. Since 1985, there have been at least a dozen more Abbott court rulings impacting NPS and the 30 additional Abbott districts around New Jersey.

The other significant policy affecting the NPS can be traced back to the July 5, 1995 vote by the New Jersey State Board of Education to accept the Education Commissioner's recommendation to take over the Newark school district. On July 12, 1995, New Jersey State Troopers escorted the then superintendent of the Newark Schools out of the central office building; with this action

Newark became the third state-run school district in New Jersey. The NPS are still, as of this writing, under state control.

Partnership with Montclair State University

For more than three decades, Montclair State University has had a presence, involvement, and commitment to the Newark Public Schools. This has manifested itself through various grants, programs, and activities throughout the years, aimed at building educator capacity in the Newark Public Schools. The President of Montclair State University, Susan A. Cole, is quoted in the latest NPS Strategic Plan regarding the University's commitment to NPS:

> All children in our society deserve the opportunity to succeed in school, and we know, unequivocally, that teachers are the most important determinant of student achievement. At Montclair State University we are committed to working in partnership with the Newark Public Schools to educate outstanding teachers who will make a difference in the lives of Newark's children and prepare them for full and productive participation in our democratic society.
>
> (Newark Public Schools, 2009, p. 6)

As a public school principal who has had preservice teachers from other colleges and universities in the Newark area, I respectfully liken the relationships and experiences to "leaving the baby on the doorstep." This is in stark contrast to the relationship and experience of having MSU preservice teachers in our schools. MSU promotes simultaneous renewal of schools and teacher education through collaboration and partnership between and among the university and their partner public school districts. Since the early 1990s, MSU faculty members from education and the arts and sciences teach courses and provide professional development and coaching to teachers on site in partner schools. Currently the MSU entity responsible for coordinating and promoting university-school partnerships is the MSU Network for Educational Renewal (MSUNER). But, long before the MSUNER began as the Clinical Schools Network, MSU had a major impact on teachers' professional development in Newark through Project THISTLE (Thinking Skills through Teaching and Learning).

Project THISTLE, funded by the Victoria Foundation, was one of the longest running and most pervasive of the NPS-MSU partnership endeavors. THISTLE was first implemented in 1979 and was in continuous operation until 2009, and included providing stimulating professional development to classroom teachers in three overlapping and sequential phases: (1) graduate coursework to improve their own understanding of the learning process, (2) additional related coursework, and (3) extended professional development activities.

There were several long-range benefits to the participants of THISTLE in addition to the benefits that accrued to the students they taught. Studies and evaluations of THISTLE consistently showed over the years that it was effective in improving students' higher order thinking skills and reading comprehension. For many of the teacher participants who completed the THISTLE graduate classes, this professional growth experience became a catalyst for them to pursue additional studies. Over 30 years, hundreds of Newark teachers took part in THISTLE and some of them became leaders in the districts in central office and building level positions.

The Harold Wilson Middle School for Professional Development

Perhaps the most ambitious and complex partnership endeavor between MSU and NPS was the Harold Wilson Middle School for Professional Development (HWMSPD). The concept of Professional Development Schools (PDS), especially the highly successful Schenley High School model in Pittsburgh, represented the cutting edge in teacher education and partnerships between schools and universities. The groundwork of partnership laid in Newark by Project THISTLE bore new fruit as the idea of a PDS in the Newark Public Schools became the new focal point for collaboration between NPS and MSU. During the 1989–1990 school year, individuals representing the university, the district, and the Newark Teachers Union (AFT) met regularly to map out all the particular dimensions of this complex school and learning community. This group later evolved into the governing and decision-making body for HWMSPD. The school's motto was, "Everyone Teaches, Everyone Learns!"

In September 1990 the doors opened to welcome the neighborhood's 6th-, 7th-, and 8th-Graders. The school was located in the Central Ward of Newark, which still showed the signs and ravages of the Newark riots of the late 1960s. Empty lots, abandoned buildings, and dilapidated housing became the backdrop for the creation of this promising new school. None of the partners on that opening day could realize how much bigger this school would become than our combined ideas, hopes, and visions in just five short years. This is perhaps best illustrated by the New Jersey Department of Education's report in early 1995 recommending state takeover of the NPS. The report cited HWMSPD as one of the few bright spots in the entire district. The NJDOE report went on to encourage the NPS-MSU partnership to continue even into the takeover.

HWMSPD provided a multitude of experiences and services simultaneously right from inception. Much of the funding to support this school came from the Abbott funding formula. Listed here are some examples:

- The cadre of teachers selected in the application process to become the middle school faculty was considered to be among Newark's very best teachers. The daily schedule dedicated significant time to this group for professional development, collaboration, and collegial exchange of talents and resources.

- Several university faculty members took up full-time residence at HWMSPD. There were additional MSU faculty members who worked part-time with the school, usually on very specific and focused areas for professional growth and planning.
- Many Montclair State teacher education students had their field experiences at Harold Wilson, including student teaching, and these graduates have made an impact on public education and partnership activities in significant ways. The current director of the MSUNER student taught at Harold Wilson as part of her MAT program at Montclair State.
- A great deal of time and attention were devoted to building strong rapport and relationships between the two faculties. This was important and critical to the success of our day-to-day work together. When the two faculties blended together, HWMSPD was at its strongest levels of performance and outcomes.
- The middle school classrooms became clinical laboratories where new professional development ideas and concepts were implemented, practiced, and honed. Every teacher from both sides of the partnership became a reflective practitioner who came to value peer coaching, team teaching, walk-throughs, collegial feedback, and collaboration.
- A full-time professional learning experience was developed for Newark teachers from other Newark schools. A system was put in place to cover 25 classrooms across the district for five full weeks. The HWMSPD and MSU faculties became the teachers and clinical laboratory for this extensive PDS experience. The goal always was to model theory into practice to reflection to refinement and improvement, thus ultimately resulting in improved student achievement.
- HWMSPD became designated as a Clinical Site for John Goodlad's National Network for Educational Renewal (NNER). In fact, the existence of the Harold Wilson partnership was a strong element of MSU's successful application to become an inaugural member of the NNER.

The Demise of the Harold Wilson PDS

Unfortunately, in June 1996, at the end of the first school year under state take-over, the staff of HWMSPD was informed that the school would close and be disbanded at the end of that same school year. This decision came despite the significant academic gains made by the middle school students and the valuable professional development services being rendered by the Harold Wilson faculty. The rationale offered for the decision was a planned shift from Harold Wilson as the primary site for teachers' professional development to a model where coaches would be in residence in each school in the district. I believe that each of us who worked at HWMSPD can tell you (to this day) exactly where we were when we heard the news that the school was closing. Disbelief, dismay, anger, sadness, and shock swept over us in constant waves for days after the news arrived. In the air at the school was a pall of profound grief for the loss of such important and successful work for the students and teachers of Newark. A year prior we had worried and wondered what effect, if any, the state takeover would have on our work and school. Now we knew.

But, fortunately, the story does not end there. In the second year of state takeover the HWMSPD staff was reassigned and dispersed throughout the entire district. A number of individuals ended up in leadership positions at central office or as school-based administrators. In almost every case they carried with them the seeds of knowledge, vision, high expectations, best practice, and a strong resolve to replicate and continue the excellent work of HWMSPD, including partnership with Montclair State. These individuals became change agents and catalysts wherever they ended up after HWMSPD, and played significant roles in creating the pockets of excellence that exist today around the NPS.

The Benjamin Franklin Elementary School

In early August 1996, I was wondering where I would end up working next, as HWMSPD was being disbanded. An invitation to apply for a vice principal position in the North Ward of Newark came to my attention. In a few short weeks I applied, was interviewed, and hired to become the new Vice Principal of Benjamin Franklin Elementary School (BFES)—one of Newark's largest bilingual schools, with a record of very low student achievement. Although I had 24 years of teaching experience in the Newark Public Schools, I knew absolutely no one in that part of the city or at this particular school site. I felt alone and lonely in my new role and new setting. Although there was no formal partnership in place specifically between BFES and MSU, I knew that I wanted to stay tethered to my partner friends—the colleagues who gladly offered me all the support, wisdom, and sage guidance I wanted. In retrospect it is highly impressive how strong the ties were between partners. I never sensed any impatience or dismissive attitudes on the part of any of my MSU colleagues; quite the opposite, they displayed great enthusiasm and encouragement for the work I was immersed in at BFES. Their offers of time, resources, and materials were sincere and generous.

Eighteen months into my administrative work, through a string of quirky circumstances, I was named the new Principal of BFES. I introduced the MSU Network for Educational Renewal to the staff, with a few staff members responding. Over the next few years I hosted a MSU faculty member full-time to work with new first, second, and third year teachers and began building classrooms that would be able to host preservice teachers as exemplars of good classrooms. Our school welcomed large numbers of MSU students to our building so that they could fulfill their obligation to spend time in an urban school setting. With the support of the partnership and the dean of the College of Education, more MSU faculty members came to the building to provide professional development or to work with small numbers of teachers. During this phase of transforming BFES, I remember feeling like a single ant moving the pile of sand one granule at a time from Point A to Point B; there were moments when I faltered and felt defeated, but the partners kept me buoyant. I became fueled and motivated by the feedback my MSU partner-colleagues were hearing back on campus about this "impressive and amazing urban school called BFES." We were helping MSU to increase their preservice applicants to work in urban settings, showing how powerful the potential of partnership is when it is tapped and allowed to flourish.

The Partnership for Instructional Excellence and Quality or PIE-Q

In 2004, a new chapter of formal partnership and collaboration between MSU-NPS was launched, called PIE-Q (Partnership for Instructional Excellence and Quality). Born of the desire to form a more effective and focused partnership between Montclair State and a small set of schools in Newark that would emulate and scale up the partnership that had developed with Benjamin Franklin School, the University applied for and won a teacher education partnership grant of $80,000 from the New Jersey Department of Education. PIE-Q links all aspects of teacher preparation and development to affect interdependence in the recruitment, preparation, retention, and professional development of urban educators so that renewal of the schools and teacher education occurs simultaneously.

A network of seven elementary and secondary schools forms the nucleus of the relationship between MSU and NPS, whereby new procedures, policies, and practices are tested for use throughout the partnership. The goal is to create a continuous cycle of growth and renewal simultaneously at the school, university, and individual teacher level. At each of the seven PIE-Q school sites around Newark we have seen success in affecting the recruitment, preparation, retention, and professional development of urban educators.

PIE-Q schools have access to partner resources and services to assist them in achieving the goals and mission of this partnership. These include:

- MSU faculty and facilities were made available to PIE-Q schools.
- Collaborative projects were encouraged by both partner constituents. Some MSU faculty took up weekly residence at the school site; they conducted their classes on site and utilized public school staff to either co-teach or make special presentations throughout the semester.
- Leaders of Teacher Preparing Schools (LTPS)—This was a multi-year, federally funded grant to the University of Washington and the NNER, which funneled money to member settings after an application process. It afforded urban schools around the nation an opportunity to strengthen and support current school principals as well as to cultivate and nurture teacher leaders. PIE-Q schools in Newark were able to send 15 administrators, district personnel, and teachers to Seattle. All of the Newark teachers who

were involved in LTPS have completed formal academic programs leading to state certification for principal and/or supervisor. Incredibly, all of them have been promoted to formal titles and roles of school leadership within the NPS. These individuals are evidence of the importance of building capacity from within rather than relying on a small cadre of existing leaders.

- Benjamin Franklin Elementary School in Newark was the recipient of the NNER's 2009 Exemplary School Partner Award.
- Four PIE-Q schools were able to achieve Adequate Yearly Progress (AYP) (as defined by the federal No Child Left Behind law) on the state assessments after multiple years of failing to reach AYP. This took them off the state list of Schools in Need of Improvement.

In June 2008 the American Association of Colleges for Teacher Education (AACTE) hosted a congressional briefing entitled "Partnerships That Work: Turning Around Low-Performing Schools." This briefing in Washington, DC highlighted exemplary models of university-public school partnerships from around the country that were having a significant impact on student academic achievement. MSU-NPS was invited to present about the work and achievements of PIE-Q. Ada Beth Cutler, Dean of the MSU College of Education and Human Services, and Jennifer Robinson, Director of the Center of Pedagogy at MSU, participated in the presentation of the program together with NPS Roger Leon, then Assistant Superintendent of Secondary Schools. They presented an overview of the scope and range of PIE-Q activities. Data showing significant growth in student achievement at PIE-Q schools since the establishment of the partnership were presented as evidence of the impact of the partnership.

The longevity and depth of the NPS-MSU partnership, especially its most recent manifestation in PIE-Q, put it in excellent standing to apply to the U.S. Department of Education in Washington, DC for an Urban Teacher Residency program under the Teacher Quality Partnership grant program. In 2009 district and university personnel worked closely with the PIE-Q governing body to complete the complex application. It was with excitement, humility, and trepidation that we celebrated in early October 2009 the news that we were one of the first awardees in the country for our proposal for the Newark-Montclair Urban Teacher Residency Program with a grant of $6.3 million for five years. The proposal received perfect scores of 100 from its reviewers and it was posted on the USDOE website as one of four exemplary applications. Once again, the intersection of new policies/programs and the existing record of partnership between MSU and the schools brought about new opportunities for success and innovation in teacher education across the continuum of teacher development. In late January 2010 the first cohort of Elementary Residents began their 18-month journey to become dual-certified Elementary/Special Education teachers. In June, the first cohort of Secondary Math and Science Residents began their

12-month journey to teacher certification and new professional development and induction program plans are underway.

Personal Reflections on Partnership

In the spring of 1990, I was invited to apply to teach at a new school that was opening in Newark—Harold Wilson Middle School for Professional Development. I was told that this would be a partner school between Newark Public Schools, the Newark Teachers' Union, and Montclair State University. I said, "No thanks." Two more invitations were extended to me to apply to this "exciting and innovative new school." Both times I declined; I did not know what a partner school was and didn't think that I needed to find out. Finally, in October 1990, I was invited again to apply and this time I said "Yes." I had absolutely no idea that I had just made a decision that would become a watershed moment in my career and my life. I refer to my time at Harold Wilson Middle School PDS as my period of personal and professional awakening.

For the first time in my teaching career I was immersed in a school culture that valued and emphasized ideas, standards, and practices that I never knew existed. Early on in my experience at Harold Wilson came a tremendous personal and professional reckoning: I was not the great teacher that I thought I was. As a matter of fact, I kept learning how much I didn't know and how much more there was to the art and science of teaching than I had ever realized. The notion of reading current research, professional books and journals was foreign to me and had never been presented to me as a professional norm in any of my prior teaching experiences. Bloom's Taxonomy, Pedagogy, Backwards Planning, the Agenda for Education in a Democracy, Developmentally Appropriate Instruction were all anomalies to me—I just did not know! But instead of being overwhelmed and daunted by all of these new and foreign facets of education I found myself excited, stimulated, and drawn to drink it all in and to gain a rich working knowledge of everything. My six years at Harold Wilson can be characterized as being filled with gaining knowledge, honing my teaching skills and craft, striving to attain levels of best practice, and interacting daily with peers and colleagues from Newark Public Schools and Montclair State University who inspired and taught me more than any formal academic program ever had.

If I had to identify one prominent learning experience that I gleaned from my years at Harold Wilson it can be summed up by a conversation the staff had with John Goodlad one day when he was visiting our school. (Harold Wilson PDS was also one of Goodlad's NNER Clinical Sites.) Goodlad was making the point that the teacher is the single most important factor in influencing the quality of a child's school experience. The quality of the teacher defined the quality of each student's education. Goodlad went on to speak that day about the fact that true school reform has to happen classroom by classroom. I can remember this session with Goodlad as clearly as if it happened today. His words summarized and climaxed the extraordinary and life-changing events of the PDS for me in a way that is difficult to explain in mere words. There is a Greek word *raima*, which means "God breathed." One connotation of this word is used to explain those "Aha" moments in your life, when an idea or concept goes from the printed word to the inner core of your soul and takes up residence. This is what happened to me during this Camelot phase of my career—I became consumed and obsessed with the importance of best teaching practice. I felt convicted for all the prior years when I thought I was doing a good job when in reality I was performing at a low, rote level.

Little did I know that this rich PDS experience would become the backbone of my years in school administration. In 1996, I was appointed as Vice Principal of Benjamin Franklin Elementary School in Newark. A year and a half later I became the Principal of this school and have worked tirelessly to provide the teachers and staff with as rich and informing of a professional experience as I possibly can. Fortunately, my Montclair State University partners have been a part of this journey to renew Benjamin Franklin School. The dimension of "partner school" is a valuable and rich component of Franklin School that has been integral in many of our accomplishments. Some examples include: student achievement test scores range in the 70 and 80 percentiles consistently compared to 12 years ago when they were single digit percentiles! The school has achieved Adequate Yearly Progress consistently and has started a committee to apply for National Blue Ribbon Status. In March 2010, we were notified by the New Jersey Department of Education that we were one of nine schools in the entire state selected to receive a Title I School Rewards Grant from federal funds for $100,000 because of our outstanding achievements the past few years. In 2009, the NNER awarded Franklin School the Dr. Richard Clark Exemplary School Partner Award, which we are extremely proud to receive.

In essence, the intersection of forward-thinking leaders at Montclair State; federal, state, and local policies and opportunities that valued and funded school-university partnerships; and my own development as an educator in the Newark Public Schools transformed my professional life and I believe the lives and futures of the students I have been responsible for in my classrooms and in the Benjamin Franklin School. I am enormously grateful for and humbled by it all.

The Partnership from the Perspective of Montclair Public Schools—Frank Alvarez

The Setting

Montclair (NJ) is a diverse community in northern New Jersey, approximately 15 miles from New York City and home to Montclair State University. A thriving suburb, the community is known for its urban flare and its economically and racially mixed neighborhoods. Its two dozen religiously affiliated institutions also speak to its diversity. Its many parks, tree-lined streets and easy commuter access to New York City make this an ideal community in which to live. However, it is Montclair's nationally renowned and racially diverse public schools that have drawn families to the community for many years.

Montclair Public Schools is a unique educational system that has received national recognition as a model for high quality, integrated public school education. The school district is comprised of eleven schools—seven elementary schools, three middle schools and one high school. It developed a system of magnet schools in which parents select the elementary school that best matches the needs of their child. Although all schools have a common core curriculum, each school delivers this curriculum through thoughtfully designed programs based on a series of themes—science and technology, global studies, liberal arts, environmental science, gifted and talented, and Montessori.

The choice system was implemented initially as a voluntary desegregation plan in 1977. Beginning with only two magnet schools, the plan has grown to include all schools. The magnet system provides parents with a choice. Parents feel empowered and become very involved in their child's learning and in supporting the specialized needs of the school. The magnet plan has had a positive impact on the community by providing alternatives within the public schools and creating exciting learning environments.

Partnership with Montclair State University

Innovation and high academic standards have been driving forces in the school district. Membership in the Montclair State University Clinical Schools Network—the opportunity to be reflective practitioners committed to teaching for critical thinking and democratic ideals, as well as to engage in best practices and institutional change—seemed only natural to district staff when first proposed in 1985.

Montclair was one of the first districts to enlist for participation in the Network. Its involvement has grown over time and has been redefined from time to time. We do not propose to do it all or to do it best. We do believe, however, that there is an honest commitment to the core values of the Network that has been expressed in our actions over time. Professional renewal opportunities, Clinical Faculty appointments, professional development school status, university-school partnerships, and service learning experiences are all part of our story.

Professional development for staff is probably the single most important element of our relationship with the Network. The ability to be a part of an ongoing community of practice that thinks deeply about what happens in schools and reflects critically on issues of teaching and learning, research-based best practices and student academic achievement is essential to teachers' professional growth and to the progress of schools in meeting the needs of students.

This professional renewal for staff happens in several ways—graduate coursework, workshop participation, mini-courses, special seminars, study groups, and

Clinical Faculty status at Montclair State University. Attendance in graduate level classes, workshops, and seminars provides exposure to a variety of topics and introduces staff to the latest innovative practices in education. This more often than not results in conversations at the school and district level, where teachers assume leadership roles in introducing best practices and in encouraging school renewal efforts.

One benefit of membership in the Montclair State University Network for Educational Renewal (MSUNER) is annual funding for Teacher Study Groups that enables teachers to take control of their own professional development by focusing on a self-selected topic of study. Several teacher-led study groups have emerged over time in the Montclair Public Schools. One example involved a group of 12 teachers who, along with a faculty member from Montclair State, met once a week for 25 weeks during the school year. *The Mindful Teacher* by Elizabeth McDonald and D. Shirley (2009) and *Mindful Teaching and Teaching Mindfulness: Guide for Anyone Who Teaches Anything* by Deborah Schoeberlein (2009) were among the texts read. The study group was open to any staff member at the school. Discussions supported emotion management, stress reduction, and mind focus. One staff member described the process as follows: "This group has become a safe haven for staff to take a deep breath, release their stress and focus the mind."

A second Teacher Study Group defined itself as a Professional Collaborative Study Group. This was another school-based core team who self-identified three areas for professional growth. Eight teachers engaged in a self-reflection process using a survey to identify three themes—motivation, differentiated instruction, and seven qualities for being highly effective. A Network grant enabled the purchase of three books that focused on the above themes. The group met twice a month during the course of the year. The group's enthusiasm and camaraderie led to a request for an extension from the Network to continue its work through the end of the next academic year. One of the goals for next year is to implement classroom strategies learned and to identify an assessment that would help evaluate personal growth.

The collegial spirit and professional cooperation emerging from these loosely structured experiences has produced, no doubt, more thoughtful and skilled teachers. Staff participants are more focused on their personal improvement and in honing teaching skills that will ultimately impact improved learning for all students.

Another feature of membership in the MSUNER was the opportunity for teachers and administrators to be appointed as Clinical Faculty at the University. Clinical faculty status is achieved by fulfilling a number of requirements at the University. More important, the University seeks to identify and support teachers who see themselves as lifelong learners, who want to support new teachers, and who recognize critical thinking as central to their work in the classroom. Clinical faculty members support the induction of new teachers at the school level and provide guidance and mentoring to preservice teachers.

In Montclair, approximately 90 teachers have achieved Clinical Faculty status. This represents approximately 12% of our teaching staff. Clinical faculty members are in all 11 of our school buildings and range from kindergarten to Grade 12 teachers. Perhaps the most rewarding element of being a Clinical Faculty is a teacher's ability to contribute to the preparation of preservice teachers. The dynamics of this relationship enable the Clinical Faculty member to provide assistance to prospective teachers, thereby contributing to the profession. In addition, the Clinical Faculty member is enhanced professionally from the experience through interactions with MSU faculty and students. Finally, the preservice teacher's experience is enriched based on the fact that the Clinical Faculty and the MSU staff are aligned in their expectations and philosophy.

Another major benefit of membership in the MSUNER has been our ability to contribute to the preparation of our own future teachers. Preservice teachers from MSU are hired by our district at a higher rate than students from other colleges and universities. Over the past three years, approximately 17% of all new teachers to our district were Montclair State University graduates—the largest percentage of new teachers from any single college or university. We find that these students are well prepared and very capable of assuming classroom responsibilities independent of others' supervision. In many instances, they have superior content knowledge, strong classroom management skills, and acute alignment to our district's goals and expectations having been part of this unique Network experience.

A PDS at Montclair High School

Over the years of our membership in the MSUNER, one of the primary strategies for advancing the mission of the MSUNER was the creation of Professional Development Schools (PDS). These partner schools agree to an expanded mission where public school and university colleagues work collaboratively to improve education for current students and future educators. PDS schools provide access to quality knowledge to all learners, provide exemplary professional development for all adults in the learning community, provide school-wide clinical experiences for preservice teachers, and engage in school-wide inquiry to improve learning for all students. The PDS model grew from Montclair State University's close work with John Goodlad's Agenda for Education in a Democracy and the National Network for Educational Renewal. The Agenda's moral dimensions of stewardship, access to knowledge, nurturing pedagogy, and democratic practice guided the partnership at all levels.

Montclair High School became a Professional Development School in the early 1990s. Under the direction of a visionary principal, with funding from the New Jersey Department of Education, strong support from the teachers' association and a vibrant partnership with the MSU College of Education and Human Services, a collaborative effort was forged to transform Montclair High School. As a structure to advance the simultaneous renewal of public schools and teacher preparation programs, the MHS-MSU Professional Development School partnership saturated the high school with energy and compelling professional development opportunities. A faculty liaison from MSU taught multiple courses at the high school and sparked faculty involvement in reading and study groups, Dodge Grant-funded action research teams, and participation in the summer Leadership Associates Program. In fact, the earliest iteration of the high school's Civics and Government Institute grew from MHS faculty participation in Leadership Associates, where CGI was first conceived and developed through a faculty inquiry project.

Ultimately, the Professional Development School partnership guided Montclair High School's work on its federally funded high school redesign project. MSU professors participated on *critical attribute* teams along with MHS faculty and administration to essentially transform MHS into several small learning communities. Montclair High School has been enriched and on many levels transformed by its experience as an MSUNER Professional Development School.

While the contributions of the PDS partnership to the success and transformation of Montclair High School were many, over time the school became a PDS in name only. In addition to the common loss of energy and commitment when an initiative or project is no longer new and exciting, there were other reasons for the fading of the PDS. First, the faculty member who served as liaison to the PDS for five years decided to change the focus of his work and no other faculty member elected to take on this assignment. The absence of a regular MSU faculty presence to work with student teachers, and coach and provide other professional development for teachers, took its toll and the level of inquiry and on-site professional development diminished. Second, the principal who was instrumental in establishing and nurturing the PDS partnership retired and the subsequent principal did not have any personal or professional commitment to the PDS. The PDS governing council ceased to meet or eventually even exist, no courses were taught on site for student teachers, and the PDS existed in name only.

Nonetheless, solid and important partnership activities continued despite the loss of many features of a true PDS. MSU teacher education students continued to fulfill their field experiences at the high school. Additionally, Montclair High School teachers continued to engage in substantive and sustained professional development experiences in partnership with the MSUNER. These experiences have become part of the culture of our school, are embedded in teachers' professional lives and continue to shape faculty discourse about teaching and learning. Most recently, MHS teachers have taught MSUNER mini-courses, participated in an MSUNER teacher incentive grant-funded reading group focused on mindfulness and teaching, mentored fieldwork students and student teachers, trained MSU fieldwork students to tutor MHS students, supervised student teachers as MSUNER Clinical Faculty, participated in a summer Writing Workshop, developed inquiry projects for and helped facilitate the summer Leadership Associates Program, and developed Dodge Grant action research projects.

The Partnership Today

Today, Montclair High School's longstanding partnership with MSU continues in important ways. Montclair High School holds a seat on the MSUNER Executive Committee, while Montclair teachers help guide the work of the MSUNER Operations Committee. One of the functions of the Operation Committee members is to coordinate the field placement of preservice teachers in district schools. Most recently, the new high school principal and a veteran faculty member from the high school spent a week in Seattle with a new MSU faculty member at the NNER Summer Symposium. Their travel and expenses were funded by the Center of Pedagogy at Montclair State, from regular budget funds allocated for professional development. We all have high hopes this collaborative experience will add to the intensity of the partnership and fuel new efforts toward simultaneous renewal of the high school and the teacher education program at MSU.

The Partnership at Bradford School

Other schools in the district have benefited from the partnership in addition to the high school. Several years ago, I made a challenge to the principals of the Montclair Public Schools to revitalize the magnet themes of each of their schools to remain relevant and innovative learning environments that would attract students from the various neighborhoods within our community. Bradford School, which sits at the northern-most point of town and at the southern tip of Montclair State University, achieved the most dramatic shift by creating an entirely new relationship through our partnership with the University.

One of the original magnet schools in Montclair, Bradford had long been defined as the fundamental magnet. Its emphasis was on providing an educational program focused on delivering a strong basic curriculum in mathematics, reading, and writing. Although the school had a good track record, shifts in the community's demographics called for an overhaul in order to make Bradford school more attractive to students and parents in a system where families choose the school that best meets their needs.

Strong leadership, additional resources from the Board of Education to hire a magnet coordinator, a capable faculty that envisioned what they could become, and immense support from the Network and the University enabled the redesign of Bradford School as the University Magnet. The result is a dynamic learning environment that supports a unique partnership between the Bradford School community and Montclair State University. Walk into Bradford any day of the week and you will see University students working in classrooms, teachers engaged in conversations with MSU faculty, and students preparing to walk to the University to experience a number of activities. MSU faculty and students facilitate weekly philosophy sessions in a number of classrooms, support instruction in math centers, organize field days, present discussions on topics of interest throughout the year, and introduce students to a variety of new learning opportunities. Similarly, Bradford students can be found on the MSU campus listening to a music recital, watching a theater performance, swimming at the health center, or simply observing nature on the University's landscaped paths.

In addition, an MSU professor from the College of Education and Human Services is assigned to Bradford School one day per week to assist classroom teachers, while providing guidance to MSU preservice teachers. In return, Bradford's experienced teachers supervise many of the university's interns. This reciprocal relationship enables greater opportunities for small group learning and differentiated instruction, while achieving a lower student-teacher ratio. The success of the Bradford School redesign has been documented in a case study by the District Management Council (Cambridge, Massachusetts). It has been featured in several local media outlets and in a video on the *Edutopia* website, and has been presented to doctoral students at the University of Pennsylvania. While never formally named a PDS, the partnership work at Bradford represents the highest aspirations and ideals of teacher education partnership.

Service Learning

Service learning is an aspect of the partnership that transcends individual school settings. Students from Montclair State University participate in service learning opportunities with our district. These opportunities contribute to both our district's ability to provide extended services to students as well as the preservice preparation of future teachers. Four community sites in Montclair provide for an after-school tutorial program that serves students who are in need of academic support. The program, housed in four different churches throughout the community, provides services to elementary school students. Approximately 80 University students work throughout the year in conjunction with district staff to address skill deficiencies in students and to assist students to attain proficiency on state-administered assessments. The University students are periodically acknowledged at public Board of Education meetings for their support of our students, as well as for their commitment to our public schools.

Service learning encourages students to employ the values, skills, and knowledge learned in the University classroom and in real-life experiences through service to schools and communities. In our setting, service learning provides both the framework and the practice necessary for preservice teachers to become successful practitioners. Simultaneously, it assists our schools to provide additional services to students while providing one-to-one or small group instruction.

I have seen firsthand the impact of service learning. First, it is a great vehicle for expanding the capacity of schools to provide additional services and resources to students. Second, it provides support to K-12 teachers in the form of teaching assistants, while connecting teachers with college faculty who supervise the preservice teachers and who are able to serve as the conduits to best practices on teaching and learning. Third, it fosters the needed connection between theory and practice for newcomers to the profession, providing essential opportunities for professional discourse, trial-and-error learning, and practical applications to learning theories. Finally, it encourages a spirit of community service that is critical to the well-being of our society.

Reflections—Frank Alvarez

Our partnership continues both to challenge and support us. The challenge stems from the exposure to new information and knowledge that the University adeptly provides. It is derived from the conversations with colleagues whose ideas intrigue us and who often cast new perspectives on familiar landscapes. It is found in classrooms where teachers struggle to reach students who yearn for answers. It is heard in the whisper of thoughts as we grapple with our personal renewal.

It is the support, however, that is provided in the partnership at all levels that makes the experience incredibly valuable—like colleagues with a common purpose. I was recently asked about the Network and how teachers and Board members view it. I thought for a moment and realized that it is so much a part of what we do and how we think about schools that it is difficult to isolate. The beliefs and values are internalized and have become very much a part of the fabric of our schools.

One of the most powerful components of the Network is the informal working dynamic that has emerged as a result of this partnership. The school district's ability to reach out at a moment's notice to University faculty for assistance is invaluable. Similarly, the University's convenient access to a public school setting has its own benefits. As an example, there was an invitation by the Superintendent of Schools to the Dean of the College of Education and Human Services to join local school administrators, primarily principals and central office personnel, in a discussion on student retention. The dean agreed and spoke to the group on current research findings, presented data, and engaged the participants in a discussion. Principals are now more cognizant of the research and the long-term deleterious effects of retention on students in some cases. The results are that school district personnel are more reflective about student retention decisions and that the school district's retention rate has dropped in the past two years.

In other instances, University personnel have assisted with major curriculum initiatives, served on high profile district committees and provided professional development for staff. Similarly, the University hosts a number of events and programs in which the district is often an active participant. As superintendent, I know I have trusted colleagues at the University who share my commitments to excellence in public education in a diverse, democratic society. That knowledge and trust have been built over years of partnership, and the value of this partnership to me personally and to my district writ large is impossible to quantify.

Conclusion

Visitors to Montclair State and its partner schools from other universities often express awe at the complexity, level of human and financial resources allocated by the University, sense of common purpose, and valuable outcomes that characterize the Montclair State University Network for Educational Renewal. They are reminded, as the reader should be, that it took at least 24 years, a great deal of hard work and talent, and a strong measure of serendipity to accomplish and construct what exists today. There are five important points or lessons about successful partnership for teacher education that resonate and emerge from the three partner perspectives in this chapter.

First, the work of partnership in the MSUNER was guided and framed by shared commitments and philosophical underpinnings encapsulated in the Agenda for Education in a Democracy. These commitments are threaded throughout the narrative from the perspectives of the university and the schools, using the same language and terminology. This did not happen by chance or without great effort. In retreats, meetings, professional development programs, and trips to Seattle to the home base of the NNER and John Goodlad, faculty and administrators from the University and the schools created common understandings about the Agenda, its meaning in their local contexts, and its role in their everyday work. Without the Agenda and the time spent unpacking it and owning it, the work of the Network could not have been as focused, coherent, and meaningful for the participants. Significantly, the time and effort invested in unpacking and discussing the Agenda for Education in a Democracy and Goodlad's postulates paid off in the development of a common language reflected in each section of this chapter and in the history of the partnership. As the former dean used to say often, "This is not values-neutral teacher education. We have a point of view and it guides everything we do."

Second, important leaders and key players had unusual longevity in their positions and they fulfilled their roles with wisdom and political savvy. As noted, during almost the entire history of the school-university partnership at Montclair State, there was one Provost who was a major supporter of teacher education and membership in the NNER. The College of Education and Human Services has had only two deans in the last 30 years—one for 20 years and the second who is in her 11th year in the role. Both ascended to the deanship from the faculty at Montclair State. The MSUNER has had three directors in the last 16 years. The first was hired from outside as a faculty member and first director before becoming dean. The second and third directors were recruited from their jobs as teachers and active Clinical Faculty members in the Network. In the case of the school district of Montclair, the current superintendent served as a principal in the district for a number of years, then became superintendent in another New Jersey school district that he brought into the MSUNER. In 2003, he returned to Montclair as superintendent. His commitment to the principles and activities of the Network has continued unabated since the early 1990s. Key individuals in the

Newark Public Schools, including the co-author from Newark, earned degrees at MSU, participated in MSU programs such as THISTLE, were faculty members at the Harold Wilson PDS, took part in NNER professional development programs in Seattle, and/or served as Clinical Faculty members at Montclair State. This kind of longevity cannot be reproduced in other settings easily, but it is an important element in the ongoing and steady development of this partnership. Conversely, as noted earlier, when leaders such as PDS principals or member district super-intendents did change, it often meant a loss of commitment and involvement in those settings.

Longevity alone is not enough. The personal and professional qualities of leaders can make or break the success of partnerships in teacher education. The dean who served for 20 years at Montclair State established the partnership and initiated membership in the NNER among many other innovations he intro-duced. He garnered resources for the work by cultivating the commitments and enhancing the knowledge of his provost and president and his arts and science colleague deans. He hired wisely and carefully when he created the Center of Pedagogy with an executive director and a position for the partnership director. School district leaders such as the co-authors of this chapter lent their support and political capital to encourage participation in the Network in their schools and districts and they spoke often and eloquently about how they had benefited personally and professionally from such participation.

Third, well-timed public policy and public and private funding for partner-ships in teacher education fueled the simultaneous renewal of Montclair State and its partner schools. They also enabled innovation and experimentation that some-times advanced the work and at other times missed the mark but taught important lessons. The primary example of the latter is the PDS movement and its concom-itant funding and policies and the story told in this chapter about Montclair State's PDSs. The Harold Wilson PDS may have been dismantled by the state of New Jersey when it took over control of Newark Public Schools, but the educa-tors who took part in that grand experiment have become leaders in the district and at the University, taking the lessons learned at Harold Wilson with them.

Clearly the most seminal development in partnership policy and funding for Montclair State and its partner school districts came with Goodlad's creation of the National Network for Educational Renewal, Montclair State's application, and selection of Montclair State as an inaugural member of the NNER. The pres-tige and visibility lent by John Goodlad's reputation and renown in education, the funding from DeWitt Wallace to the NNER that was funneled to member settings, the professional development programs offered by the NNER in Seattle and in local settings, and of course the grounding provided by the Agenda for Education in a Democracy all accrued incredible value for the University and its partner schools.

Fourth, while members of any partnership should be willing to do things for the common good or even for the benefit of other partners solely, they must have their own needs met to a significant degree to sustain the overall partnership. Within the Montclair State University Network for Educational Renewal, the University gained high quality sites for its students' field experiences, the valuable input and wisdom of practice of school partners into the renewal of teacher education and its related activities, and the ability to play a significant role in the public schools surrounding the campus. School partners gained access to a rich and stimulating array of professional development opportunities for staff members that improved student achievement, the opportunity to "grow your own" teachers, consultation and partnership with national scholars and local colleagues alike, and funding for otherwise unavailable programs and activities. In the end, all parties gave and all parties got, and their differing needs and priorities blurred and blended. And, the policy surround was enabling and supportive, despite bumps in the road locally and nationally. It has been partnership at its best.

Fifth, and last, are the intangible and difficult to measure human qualities and relationships that come through clearly in this chapter from all partners. It is impossible to miss the passionate dedication to high quality teaching and learning that flows to, from, and among the participants in this endeavor. Furthermore, the trust built up over many years of shared toil in the fields of education and partnership work is manifest in the voices of the co-authors. Despite moments of disappointment and some difficult events, the partners persisted and continued to turn to one another for help, learning, support, and even validation. Without trust and shared commitments, it is difficult to sustain partnership over time and to weather events that can sabotage or even destroy it. Clearly this partnership had that trust and shared commitments.

Partnership is now so deeply embedded in the fabric of teacher education at Montclair State that is reasonable to posit it will survive future changes in leadership, new directions in education policy and funding, and the vagaries of time. What is assured is the impact of more than 20 years of partnership on Montclair State, its partner schools, their faculty and administrators, and the many thousands of public school and university students who have benefited from the partnership.

References

Goodlad, J. (1990). *Teachers for our nation's schools.* San Francisco: Jossey-Bass.

Goodlad, J. (1994). *Educational renewal: Better teachers, better schools.* San Francisco: Jossey-Bass.

Holmes Group. (1986). *Tomorrow's teachers: A report of the Holmes Group.* Michigan: The Holmes Group.

Klagholtz, L. (2000). *Growing better teachers in the Garden State.* Washington, DC: Thomas B. Fordham Foundation.

MacDonald, E., & Shirley, D. (Eds.) (2009). *The mindful teacher.* New York: Teachers College Press.

National Commission on Excellence in Education. (1983). *A nation at risk: The imperative for educational reform: a report to the nation and the Secretary of Education, United States Department of Education.* Washington, DC: The Commission.

Newark Public Schools. (2009). *Great expectations: 2009–13 strategic plan*. Newark, NJ: Newark Public Schools.

Patterson, R., Michelli, N., & Pacheco, A. (1999). *Centers of pedagogy: New structures for educational renewal*. San Francisco: Jossey-Bass.

Schoeberlein, D. (2009). *Mindful teaching and teaching mindfulness: Guide for anyone who teaches anything*. Boston MA: Wisdom Publishers.

EDITORS' COMMENTARY

This chapter, with its three complementary perspectives from the university and school districts, illustrates a strong partnership between a university and collaborating schools. Partnership is often called for in the literature on teacher education, and sometimes even in policy. A few states call for collaboration as part of their regulations for teacher education, but it is often unenforced. Collaboration is part of the standards for accreditation but not seen as one of the major factors for a successful bid for accreditation.

In the case illustrated here, the institution did not have a state policy base for undertaking the work, but neither were there policies to inhibit the work. State regulations in New Jersey were not onerous and detailed. It was possible, for example, to infuse critical thinking and democratic practice as themes even within the context of a limited professional sequence.

One aspect of this partnership, also embedded in some state and accreditation policy, is the appropriate role for the arts and science faculty in teacher education. As can be seen here, the arts and science faculty are seen as one of three entities responsible for teacher education—along with education faculty and school faculty. The institution used a structure, a Center of Pedagogy, as the vehicle to bring the groups together. Of course, there was no state policy mandating the depth of involvement we see here—or the idea of using a structure. Both of those ideas seem to have come largely from the involvement of the university in the National Network for Educational Renewal, as did the development of the theme of "education in and for a democracy." In this case the pressure from an external, national body assumed the form of policy and was used to accomplish the same. Involving university faculty was necessary to remain in good standing in the Network. In a sense, this was the "policy" of the Network and is an example of policy can be used to accomplish a complicated goal.

Two other aspects of policy might be noted, both local policies. Under the leadership of the provost, faculty were encouraged to engage in research, including applied research that might take place in schools, museums, or other public agencies for one-quarter of their time. This program opened the door for participation of many who were initially reluctant to participate, especially university faculty from the arts and sciences. Secondly, the university negotiated the possibility of a joint appointment not only with another academic partner, but also with an academic center such as the Center of Pedagogy. Although little used, the possibility legitimated the Center as an academic unit. One state in particular, Georgia, has taken the step of requiring that colleges and universities give credit in personnel actions for faculty in arts and sciences who support P-12 schools.

The chapter makes the point about longevity. Two deans over 30 years, both with a common vision for teacher education at the institution, are part of this case. The same provost over 20 years of that time who understood the work played a critical role. These are not factors that can be controlled. Deep change with or without policy support is a long process and if it is to outlast leaders, it must be deeply embedded in the culture. On the other hand, it cannot be so deeply embedded that it is immune to change. The successful pursuit of the Teacher Quality Partnership Grant discussed in this chapter is an example of moving forward based on past success, but a move that is likely to bring change.

Discussion Questions

1. Is it a paradox that most states do not expect collaboration between universities and school districts in policy, but expect that a P-16 system will emerge?
2. Is it a paradox that arts and science faculty are often reluctant to be involved with P-12 schools in part because of reward systems that tend to focus on their narrow research interests, while the quality of future students at a university depends on deep content knowledge on the part of teachers?
3. Is longevity always a good thing? How can one use policy to intervene if longevity is a problem rather than a positive element in a given circumstance?
4. Do you think the state has the authority to require policy at universities to reward faculty for school work in decisions about reappointment, tenure, and promotion?
5. Does the use of external organizations, illustrated here by the involvement with the National Network for Educational Renewal, have the same effect as formal state policy in bringing about change? Is it possible that it has had a greater effect?
6. Can the absence of policy be a force for change?
7. Several examples of local policy that supported change are illustrated here. Which is more important in bringing about change—local, state, or federal policy? How does the impact differ depending on the nature of the change?

8

UNITED WE STAND

Divided We Fail Our Communities and Hence the Public Good

Van Dempsey and Deborah Shanley

One of the great challenges of any policy initiative is to structure its intent in such a way that its positive effects are maximized and any unintended negative effects are limited. This is particularly true for federal policy related to education. No public responsibility is more ubiquitous than education, or reaches more people in more places in as varied a range of challenges and conditions for implementation. Any federal policy related to education must be implemented across 50 states and the District of Columbia, at least in the public sector, and must then reside in its implementation in state contexts that vary as much within some states as they do from one state to another. This process is compounded by the very nature of policymaking and implementation itself: it is inherently, and by design, a political process. As Noblit and Dempsey (1996) described the policy process in the mid-1990s, "Educational reform has become a national pastime. Politicians have found that there are careers to be made in attacking and making policy to reform schools" (1996, p. ix). One and a half decades later, little has changed; if anything, the pace is more intense, and the reach of policy from the federal level to the state and local levels more intense and more constant. Noblit and Dempsey offer this critique of the process:

> Educational policy has become a way for politicians and government offi-
> cials to control the public—to make schools do what is politically desirable.
> However, it is clear from decades of research that educational organizations
> are unique. They are not like businesses or the armed forces, for they must
> represent a host of constituencies with conflicting views about what educa-
> tion ought to be doing.

(1996, p. 202)

And in calling for an end to educational policymaking as we know it, they reframe the process in this way:

> The end of educational policy would ask politicians and government offi-
> cials to make substantive decisions. It would require them to ask, "Does this
> legislation or regulation promote citizens and educators participating in
> decisions about what is to be valued in their schools."
>
> (1996, p. 203)

This chapter examines education policy in two different contexts: one urban and one rural. We analyze how a one-size-fits-all policy impacts these two kinds of communities and the fit of national policy mandates and directives in these two contexts. The first case focuses on New York City and is told from the perspective of the Dean of Education at Brooklyn College. The policy in question relates to the state's process for approving new academic programs and providers of certification and, more specifically, as part of pursuing Race to the Top funding, letting the Board of Regents allow providers outside of higher education to grant master's degrees when the program of study can lead to teacher certification. This first portrait discusses the basic policy driving the change in the state and its impact on teacher education programs. The analysis also will be framed around issues unique to this urban context that help to understand the complexities and nuances of policy development and implementation as they play out in urban places. The second portrait focuses on how federal Title II grants (and, as a matter of policy context, Title I of the Elementary and Secondary Education Act) intended to support teacher quality through K-16 partnerships impact the ability of rural schools and institutions of higher education in West Virginia to improve teacher practice, teacher education, and student learning. This portrait describes how broad definitions of poverty actually limit the ability of institutions in one of the most economically challenged states in the country to access funds that supposedly target such situations. The chapter will close with similari-ties and differences that emerge from a discussion of these two policy initiatives in these two contexts, what urban and rural places have in common as they attempt to implement federal policy impacting teacher education, and how they differ.

New York State, Program Approval, and Alternative Licensure

An Overview of Contextual Factors

The New York City public school system serves over 1.1 million students, with more than 135,000 people working full time, of which 80,000 are teachers in over 1,600 schools. In 2009, 65% of the new hires were in three areas: science, special education, and speech improvement. Identifying highly effective teachers for high-need schools continues to be a focus to address achievement gaps that are disturbing and unacceptable. The current Kids Count data reports 27% of children living in poverty, 43% living in single parent homes, and 47% of children speaking a language other than English at home in New York City. Additionally in 2008, 27% of New York City children were living in households where no parent had full-time, year-around employment. One academic indicator illustrates the depth of the problem on one required 4th-Grade assessment in New York State: 72% of fourth graders scored below proficient in reading this past year (2009–2010). By some measures the graduation rates for African American males is around 35%.

Two recent policy actions in the State of New York help frame the discussion for how to provide excellent teachers who can be successful in the context described above at levels to meet the employment demands for the city's schools. The first, in August 2010, related to the general process for proposing new institutions of higher education to operate in the state, opening new campuses of existing institutions, and adding new programs of study. The specific policy issue was the Board of Regents' role in the coordination of higher education, in particular the Regents':

- Review of master plan amendment proposals to establish new higher education institutions and of proposals by New York higher education institutions for approval of major changes in mission (e.g., opening a branch campus, moving to a different level of study).
- Review of requests for permission to operate in New York State from out-of-state institutions; and
- Authority with regard to higher education available to New Yorkers online.

The second, in April 2010, involved establishing pilot programs for clinically intensive preparation routes for new teachers, including those offered by providers outside of higher education. This component, as in many other states, parallels efforts to align state policy with efforts to secure federal Race to the Top funding.

New York Master Plans

Education law in the State of New York provides that:

> No individual, association, partnership or corporation not holding university, college or other degree conferring powers by special charter from the legislature of this state or from the regents, shall confer any degree or use, advertise or transact business under the name university or college, or any name, title or descriptive material indicating or tending to imply that said individual, association, partnership or corporation conducts, carries on, or is a school of law, medicine, dentistry, pharmacy, veterinary medicine, nursing,

optometry, podiatry, architecture or engineering, unless the right to do so shall have been granted by the regents in writing under their seal.

(NYSED/USNY, 2010, §224[1])

This provision applies to both New York degree-granting institutions and to higher education institutions located outside the State that wish to operate in New York. State education law also directs the Regents to promulgate a master plan for higher education, which is called the Regents Statewide Plan for Higher Education:

> The regents shall, on or before the twenty-fifth day of April nineteen hundred seventy-one and each fourth year thereafter, request the state university trustees, the board of higher education of the city of New York [now the City University Board of Trustees], and all independent higher educational institutions to submit long-range master plans for their development. Such request shall specify the nature of the information, plans and recommendations to be submitted, shall describe statewide needs, problems, societal conditions and interests of the citizens and discuss their priorities, and provide appropriate information which may be useful in the formulation of such plans.
>
> (NYSED/USNY, §237[2])

The Plan incorporates the State University of New York (SUNY) and City University of New York (CUNY) long-range plans, as well as the plans the Regents request from the independent and proprietary institutions and sectors. The Regents are perhaps the most powerful governing body for education in the country with authority over P-12 schools, public and private higher education, museums, libraries, and a number of professional licenses. The SUNY and CUNY long-range plans and any revisions thereof, are subject to the review and approval of the Regents and the Governor. It is the principal document guiding the New York Department of Education (hereafter referred to as Department) in higher education. In addition, Regents and Department legislative and budgetary priorities in higher education, as well as the shape of provisions in the Rules of the Board of Regents and the Regulations of the Commissioner of Education relating to higher education also reflect the Plan's priorities.

In relation to the Statewide Plan, the Commissioner's Regulations require that "to be registered every new curriculum shall be consistent with the Regents Statewide Plan for the Development of Postsecondary Education" (§52.1[c]). The process of bringing a new program that constitutes a change in an institution's mission into consistency with the Statewide Plan is called master plan amendment. A separate provision in the Regulations requires institutions to receive master plan amendment approval for the establishment of each branch campus. Specifics of the types of programs that constitute major changes in mission and need master plan amendment approval are set forth in Department guidelines.

Circumstances Requiring Master Plan Amendment Approval

In 1995, the Legislature directed the Regents to streamline master plan amendments. In consultation with representatives from the four sectors, staff identified requirements that could be dropped or combined, to which the sectors agreed and which the Board approved. Under the guidelines in effect since then, the following types of actions require Regents master plan amendment approval as major changes in an institution's academic mission:

- Establishing a new degree-granting institution (including a permanent New York campus of a "national" higher education institution) and the degree programs it would offer;
- Establishing a new branch campus of an existing institution and the degree programs it would offer at that location;
- Authorizing the first degree program at each of five levels of study in each of ten subject areas of the New York State Taxonomy of Academic Programs. The five levels are:
  o Associate degree
  o Baccalaureate degree
  o First professional degree (e.g., J.D., M.Div.)
  o Master's degree
  o Doctorate degree

The ten subject areas are:

o Agriculture	o Fine Arts
o Biological Sciences	o Health Professions
o Business	o Humanities
o Education	o Physical Sciences
o Engineering	o Social Sciences

Contents of a Proposal Requiring Master Plan Amendment Approval

A proposal that requires master plan amendment approval must provide both:

1. The academic information needed about the curriculum, faculty, academic resources, and admission and other academic policies in order to determine whether it meets the quality standards for registration; and
2. Planning information needed to determine
   a. the need for the program(s),
   b. the program(s) potential effect on the institution, and
   c. the program(s) potential effect on other institutions in the region.

The Department evaluates the academic information against the quality standards in the Commissioner's Regulations that all programs proposed for registration must meet. The academic information may include evaluations by external experts in the discipline (required for graduate level programs).

a. *Need for the Program* as a function of (1) student demand, (2) that of potential employers (local, regional, state, national, international) or (3) societal need, or (4) mission relevance to the institution;
b. *Potential Effect on the Institution* in terms of enrollment, revenue, and expenditures; and,
c. *Potential Effect on Other Institutions* with similar programs, including canvassing of degree-granting institutions in the region in which the program(s) would be offered, to give them an opportunity to comment on their perception of the need, the extent to which their own programs address that need, and the effect of the proposed programs on their own programs.

Canvassing institutions was not the Regents' preferred procedure to assess proposed programs' effect on other institutions. The 1972 Statewide Plan called for regionalization "for maximum efficiency"; in each new Regents Postsecondary Education Regions, the Board would establish a Regents regional advisory council composed of all the region's higher education institutions. The plan also called for public representatives to (1) assess regional needs, (2) inventory regional resources in terms of facilities, faculty, educational programs, and unused capacity, (3) determine "the appropriate roles and levels of participation by private and public institutions in meeting the total needs of the region," and (4) attain agreements "among institutions in regard to areas of academic program specialization with appropriate consideration of regional needs, the relative strengths of the institutions, and the views of various interest groups of the region." Councils were established in four regions. In those regions, the Department referred proposed master plan amendments to the council for review and advice. In the other regions, the Department directly canvassed institutions until councils could be established. However, no further councils were established and, as the existing councils withered, the Department extended the canvass process to the regions they formerly served. Today, no Regents regional advisory councils exist and the Department canvasses institutions in all Regents Postsecondary Education Regions.

Permission to Operate in New York State

New York degree-granting institutions have the authority to operate at their authorized locations within their approved academic missions; however, any operation in the State by an out-of-state institution requires the Board's prior permission. Regents' policy has been to limit approval to specific periods, usually not exceeding five years, with the possibility of renewal for additional terms.

A concern expressed by New York institutions was that Department reviews of out-of-state institutions do not match those of New York institutions in terms of comprehensiveness and rigor. However, all institutions, whether in-state or out-of-state, seeking to offer instruction in New York are supposed to meet the same quality standards which are those set forth in the Commissioner's Regulations for program registration and, in the licensed professions, statutory provisions for students practicing the profession. Out-of-state institutions must also provide evidence of need for their activity, in the same terms as discussed above under Master Plan Amendment.

Generally, out-of-state institutions make three types of requests to operate:

1. Offer full degree programs. Proposals by out-of-state institutions to offer full programs in New York receive the same type of review as a proposal from a New York institution that would require master plan amendment approval.
2. Offer a limited number of credit-bearing courses, without offering a full program. Such proposals receive the same type of review as a proposal from a New York institution to open an extension center (requiring the Commissioner's approval). The content differs in some details. The review often includes a visit to the proposed location as well as a canvass of institutions in the region.
3. Make use of clinical or other facilities for the education of its own students. This may include out-of-state institutions offering only online courses when their students seek placement in facilities in New York for practica and internships.

Unlike a New York institution's statutory right to a hearing on a proposed master plan amendment, the Board has the discretion to hold such a hearing on an out-of-state institution's application to operate.

Online Education

Members of the Board had expressed concern about the educational experiences of New Yorkers taking online education from non-New York institutions. Oversight differs from state to state and accreditation requirements differ among accrediting agencies. In March, the Higher Education Committee discussed online education, primarily in terms of New York institutions. The discussion paper for the committee's consideration of online education in March 2010 provided the Statewide Plan's statement about distance education, background on online education in higher education in New York and nationally, and information on good and poor practices and on reviews of institutional capability to undertake study online. Because a growing number of institutions want to place students in New York facilities, Regents approval would be required for them to operate.

Clinically Intensive Teacher Preparation Routes through Alternative Providers

In April 2010 the *New York Times* reported that the New York Board of Regents would review a proposal to create new routes to master's degrees for candidates interested in graduate study and teacher certification. The *Times* article stated:

Under the pilot, the Board of Regents will invite groups like Teach for America to create their own master's programs. The programs would need to have a strong emphasis on practical teaching skills, a nod to criticisms that traditional education schools spend too much time on theory. The Board of Regents would actually award the degree to the teacher, who would commit to a high-needs school for four years.

(Foderaro, 2010)

In an April 19 memo to the Higher Education Committee of the Regents, the committee was asked to consider for adoption at that day's meeting "Emergency Adoption of Proposed Regulations Relating to the Establishment of Graduate Level Clinically Rich Teacher Preparation Pilot Programs" (subject of the committee memo). The memo states in part:

The proposed amendment establishes two tracks for the graduate level clinically rich program: 1) the Model A track is the residency program for candidates working with a teacher of record in a high need school; and 2) the Model B track is the residency program for candidates employed as teachers of record in a high need school who will be eligible to receive a Transitional B certificate upon completion of required introductory preparation, tests, and workshops. To qualify for an initial teaching certificate in New York State, candidates in both tracks must pass the New York State Teacher Certification Examination (NYSTCE) required for the certification title. Candidates who are employed as teachers of record (Model B track) shall be eligible for the Transitional B certification upon showing evidence of completing an introductory component and passing the required NYSTCE(s) for the Transitional B certification. The proposed amendment authorizes institutions, other than institutions of higher education, and that are selected by the Board of Regents, to offer the Model A and Model B tracks of this program. Such institutions shall include, but not be limited to, cultural institutions, libraries, research centers, and other organizations with an educational mission that are selected by the Board of Regents for participation in this pilot program through the RFP process.

(New York State Education Department, 2010)

How Policymakers Came Together with the Field

Within its statutory authority, the Board has sought, since at least the early 1970s, to assure that higher education institutions operating in New York State, whether in-state institutions or those from outside the state offer instruction that meets New York's established quality standards and for which a need has been demonstrated. In doing so, the Board balanced concerns about competition among institutions and access to higher education. Both are valid issues that need careful consideration when the Board acts on major changes in the mission of New York institutions or on applications by out-of-state institutions to operate here, on ground or online.

Living and working in the largest urban public school district in the world presents opportunities and challenges where only the strong survive in getting one of the most important jobs done—preparing highly effective teachers. Building and sustaining the conditions to meet standards set by the New York State Board of Regents in 1999 was the first of many major steps taken over a decade to improve teaching and learning in PreK-16 education in New York City. For three years, the Board of Regents, members of the NYS Education Department, deans and faculty from the IHEs, partners in the PreK-12 schools and other interested members of the public gathered at meetings around NYS to discuss a range of ways to promote the required standards.

The new standards described in Subdivision 52.12(b) of the Regulations of the Commissioner of Education also encouraged the IHEs "to develop intensive, streamlined teacher preparation programs for career changers and others holding graduate academic or professional degrees." This one sentence was the first public statement for what was to be the support needed by Chancellor Harold Levy to create the New York City Teaching Fellows program to meet anticipated vacancies in the lowest performing schools. His hand had been forced by a lawsuit by then Commissioner Richard Mills against Chancellor Levy, who also happened to be a former Regent. The action forced the city to only hire certified teachers for struggling schools and anticipated the requirements of NCLB by several years. The discussion extended the vision from the Teach for America program and was a partnership between the schools and the local colleges and universities that wanted to be part of this effort. The idea was to create a tripartite model including the NYC DOE, the schools of education including the liberal arts and science faculty, and the local teachers union (UFT). Together they created a new framework that included a different curriculum to prepare some of "the best and brightest." Now, in its tenth year, the Teaching Fellows are embedded in New York City schools and the shared data system and research studies conducted across the grades and within the content areas have informed curricular changes and expanded the work to involve other informal partners including, but not limited to, the American Museum of Natural History. More online classes to enrich science content required to meet the NYS learning standards grew steadily. Despite what might be seen as a success, the extremely tight timeline forced by the lawsuit led to some unintended consequences. For example, because initially certified teachers only were required for failing schools, some principals from schools not classified as failing advised preservice teachers about to graduate to defer certification, saying, "If you are certified, I can't hire you!" The politics of teacher education policy is very complex and not subject to fast track actions that are not thought through (Michelli, 1995).

Alternate routes, like Teacher for America and the NYC Teaching Fellows program, have not solved the problem of how to retain teachers. They are young, well-educated idealists with a "missionary zeal." But their counterparts graduating from university-based programs are leaving within five years. As Cloud reports in *Time*, the focus of the conversation has shifted to how to assess teachers so only the strong survive. No one has argued that teachers lack the content knowledge, skillset, and dispositions to structure their classrooms for academic success. It is the how to get there that is in question and no one seems to have the answer.

In New York State the Board of Regents 1999 regulations focused the work in programs preparing classroom teachers on all the right indicators for success. All the major components were in place and a reasonable timeline to achieve accreditation also (December 2004), required for the first time of all New York higher education institutions preparing teachers. Everyone went back to the drawing board, realigned and recreated programs to align with the new regulations. College presidents joined in their support by articulating the first of several important standards, that liberal arts and science faculty must play a major role in the crafting of the 30-credit-content major required for all teachers. This college-wide commitment led to overdue conversations with content area faculty to align the coursework with the State Learning Standards and begin to collaborate with the education faculty in local schools and with parents and other community members.

Although the New York State Regents knew many of the campuses were preparing teachers in clinically rich environments, they recognized the policy had to build capacity across settings by mandating specific guidelines for field experiences, student teaching, and practica. In fact, the pedagogical core requirement stated: "the program shall include at least 100 clock hours of field experiences related to coursework prior to student teaching or practica" based on grade level and other special groups to be served (see NYSED/USNY, 2010/).

Brooklyn College extended the clinically rich discussions by participating in the conversations with the Carnegie Corporation's Teachers for a New Era Project (www.carnegie.org) as part of a supporting network approach; the National Commission on Teaching and America's Future (NCTAF) that fostered a series of national conversations and encouraged state partnerships "to create new policies and practices for dramatically improving the quality of teaching" (Fraser, 2007); the National Network for Educational Renewal (NNER) that engages us in renewing conversations about the public purpose of schools; and the American Association of Colleges for Teacher Education (AACTE) that recently released "The Clinical Preparation of Teachers: A Policy Brief" (2010).

Everyone was moving toward a mutually agreed upon goal recognizing all the factors that have facilitated or hindered federal (and state) policy that, as Cross (2004) describes, make a difference: the inevitable link between social issues and education; practice and policy are based on anecdotes; presidential support helps (makes a huge difference); leadership matters; the role of the Department of Education, members of Congress and local politicians; etc. What happened within the NYC teacher education efforts?

Another Very Short Timeline

In spring 2010, in preparation for New York's Race to the Top application, the Chancellor of the Regents brought together the City University of New York's (CUNY) Vice Chancellor of Academic Affairs, the University Dean of Teacher Education, and all the Deans and/or Chairpersons who lead Schools or Departments of Education at the City University of New York campuses. The topic was to introduce the idea of extending to entities outside Institutions of Higher Education, the opportunity to offer graduate-level clinically rich teacher preparation pilot programs (the same proposed regulation is on the table for Graduate Leadership Preparation effective October 6, 2010). Why? It would add points to New York State's Race to the Top application, which was in fact successful providing $700 million for implementation. The substantial amount of the grant aside, some educators came to call it "Race to the Till," because it forced the adoption of any number of significant policy initiatives with little or no consultation or discussion to secure funding.

A discussion about the proposal to expand graduate credit-granting authority provided the opportunity for all to discuss the positive and negative sides of the proposed amendment. The main concern was in two of the areas targeted, math and science, and the lack of availability of content-specialty faculty in the liberal arts and science and/or partner with content expertise. Educators believed the soul of higher education was now at the heart of the work and to move in another direction sent them backward. Another major issue was one of transparency and equity. Would the applicants be held to the same national and state standards as existing programs were, and required to align their curriculum with the NYS Learning Standards to ensure cohesiveness in efforts? There were concerns about the costs of a residency model and that without resources it could not be sustained over time. There also was confusion over the shortage areas identified, since many graduates were already finding it difficult to find jobs in math and science due to a lack of vacancies. In fact, except for Earth Science, most of the students in the city market were told to be available to substitute until the cycle changed or to pursue an extension in teaching students with disabilities or English language learners.

The meeting ended and university educators trusted that their voices were heard and that changes would be implemented into the Request for Proposals application (RFP) and that it would be circulated prior to its release for feedback. At about the same time, the Regents received a report on improving teacher education for urban schools prepared by a group of some 30 individuals representing every sector and with some suggestions for clinically rich programs. The group, appointed by the Regents, met five times and prepared a comprehensive report. That report apparently was never discussed by the Regents, at least not in public.

As many best intentions, the measure was buried in the spring/summer workings of the Regents business, making it impossible to put a proposal together unless the institution already had been awarded monies to support a residency model or was lucky enough to have a cohort group of students with a job in targeted schools. It would be impossible to recruit so quickly "the highest caliber of candidates" and set up a thoughtful, well-developed mentoring program with such a short turnaround. One element that was difficult for colleges and universities to understand was the requirement of national accreditation of all institutions offering programs for teachers with no evidence that those standards would apply to the non-collegiate programs designed to gain RTTT points.

The Teach for America organization and some of the charter school management organizations were already partnering with campuses but looking for ways to sever the ties. The tensions between highly process-driven campuses that engage faculty in the development and approval procedures were seen as cumbersome and filled with barriers to innovative learning. To operate outside of this was a key advocacy goal, especially for the public universities. Also, on the philosophical end, the need for additional study in the liberal arts and sciences for grades 5–12 were not seen as necessary. Many of our campuses were already working closely with an array of arts organizations, science museums, the national parks, and other partners.

Those developing alternate routes in New York City, including the Teaching Fellows and Teach for America, were able to use NCLB's requirement of certified teachers to expand, and in fact they could not have met the mandate without alternate routes. In New York, those joining alternate route programs attended colleges and universities to earn masters degrees during their first years of service. While the public univerisites were struggling to keep up with technology, the proponents of further expansion of alternate routes had great access to technology and used it to address the demand for accountability, sometimes in an effort to promote the alternatives over traditional programs. Fraser (2007) reflected David Imig's remarks in 2005 to the AACTE membership:

> Accountability demands will escalate ... challenges to the social justice agenda and multicultural ideology will intensify ... Calls for new forms of scholarship will escalate ... Competition among providers will intensify as school districts bid teacher preparation and principal preparation among all providers and gain the right to award certification at the district level.
>
> (2007, p. 319)

That competition was already underway in 2000 in New York City. At its monthly meetings, faculty at Brooklyn College, while working with the Teaching Fellows, reviewed research findings and presented alternate views on meeting the needs of all children but did not stand in opposition to the alternatives. It was a time when the faculty and school partners came together as they never had before, and perhaps not since, in developing new ways to meet children's needs.

West Virginia

For purposes of understanding the federal education polices to be discussed later in this chapter as they apply to rural America, it is important to understand the social and demographic contexts of West Virginia that the policy discussion will reflect. For most people who know the state, images include symbols of rustic lifestyles, and portraiture that captures the scenic beauty of the state's mountains, rivers, and valleys. Many outside the state conjure up economic images framed by coal and other extractive industries located in rural parts of the country. The state, given its resource challenges, has faced chronic problems related to poverty, and the impact that has on education and other social institutions tied to quality of life. As Dempsey notes in an examination of social class and schooling this creates:

> These key themes undergird the more general themes ... problems of access to public institutions for citizens in poverty and the working class. Even where they are available at all, geography, local politics and generally lower levels of public funding create situations where support taken for granted in some areas may not be assumed resources for people in rural places. These resources, more readily available and accessible in other places, may not close the gap left by parents', families' and communities' inability to provide them.
>
> (2007, p. 291)

He goes on to describe the context, citing Dehaan and Deal (2001, p. 53):

> While educational researchers have long attended to the many consequences of urban poverty for children, educators may be less aware that rural children face many of the same barriers and problems. More poor children live in rural areas than in cities. Rural children are likely to remain in poverty for longer periods of time, and rural children attain lower levels of education. Rural children experience greater levels of hunger, have more health related issues, and are more likely to experience substance abuse than urban children. Rural children are more likely to be involved in criminal activity. Finally, rural children are more likely to be depressed and experience chronic loneliness.
>
> (Dempsey, 2007, p. 293)

"Kids Count," the annual report from the Annie E. Casey Foundation, further illuminates this context. The 2010 report shows West Virginia as having a child poverty rate of 23%, or nearly one in four. This compares to a national average of 18%. The data is based on a definition of poverty as a family of four with an income below $21,834 (Annie E. Casey Foundation, 2010).

It is important to note that many indicators that relate to economic and resource challenges stand in contrast to investments West Virginia makes in public education. In the annual "Quality Counts" report from *Education Week*, West Virginia consistently scores low on economic factors related to "Chance for Success" (e.g. "family income," "parent education," and "parental employment," "adult educational attainment," "annual income"). On the other hand, in indicators that relate to school finance, the state ranks high in terms of providing resources—relative to available resources—to invest in public education. In the 2010 report, West Virginia ranked fifth nationally in "Spending on education— State expenditures on K–12 schooling as a percent of state taxable resources" (*Education Week*, 2010).

As noted annually in *Education Week* reports, and as will be described below, West Virginians, as much as any state citizenry in the United States can, provide for the education of its children. Further information follows on the relationship between the context of poverty in West Virginia, the specific language of Title II teacher quality policy and funding, and the impact of that policy in West Virginia school districts, schools, and institutions of higher education (IHE).

School-University Partnerships as a Promising Practice in West Virginia

West Virginia has experienced considerable success as a state at work to establish, support, and sustain School-University Partnerships, with teacher education as a central feature of this collaborative effort. This effort began in earnest in the mid-1980s when John Goodlad served as a consultant to a strategic planning initiative at West Virginia University (WVU). As part of a broad set of recommendations from multiple consultants across academic areas, Goodlad recommended that WVU invest energy and resources in the creation of a school university partnership to support the simultaneous renewal of educator preparation and professional development. The University, with the support of the Claude Worthington Benedum Foundation, undertook the initiative in 1990 to create the "Benedum Project" as a response to Goodlad's challenge and as an effort to build what was then one of only a handful of school-university partnerships in the country. The Benedum Project would be one of three goals engaged in the collaborative agreement between the foundation and the University, including:

1. To redesign teacher education collaboratively with public schools partners, involving faculty in Education and the arts and sciences;
2. To establish a network of Professional Development Schools as sites of best practice to support educator preparation and professional development; and,
3. To establish collaborative strategies to make these changes last, and to transport them across the State of West Virginia.

By the early years of the 21st century, the Benedum Project had become a solidly grounded initiative by then known as the "Benedum Collaborative." The partnership included 28 professional development school (PDS) sites, and was the working home of the partnership's Five-Year Teacher Education program. The program, a central goal and feature of the partnership, graduated its first cohort in 2000. The Benedum Collaborative would be recognized in 2003 by the U.S. Congress as an innovative partnership and Teacher Education program in the U.S. Secretary of Education's *Report to Congress on the Quality of Teacher Preparation*. The partnership was twice invited by NCTAF to participate in national summits on education partnerships, and in 2005 was invited to join the Carnegie Learning Initiative. By 2010, the partnership had secured over $4 million in private foundation and legislative funding, and a $1.7 million U.S. Department of Education technology integration grant.

In 2001, efforts began to strategically pursue the third goal cited above in the original collaboration between the Benedum Foundation and WVU: To transport the successes of the Benedum Collaboration to other higher education institutions and public schools who could work in partnerships based in simultaneous renewal. With the support of the foundation, the Office of the Governor, the West Virginia Legislature, the Department of Education and the Arts, and the Benedum Collaborative, the state embarked on an effort to create partnerships involving the state's ten colleges and universities. To guide the work, we created an advisory board comprised of tripartite representatives (P-12, education, arts, and sciences) from each local group of partners, representatives from key state agencies and the Governor's Office, and the business and corporate community across the state. At the initial convening of the potential partnerships, representatives from all ten agreed to form a school-university partnership and engage in preliminary planning for at least a three-year period. The key stakeholders also agreed to provide organizational and political support to the start-up process. In the initial year the Benedum Foundation provided a combination of pilot and operational funds for the network to support both the Benedum Collaborative and the emerging network of partnerships. The statewide network that emerged came to be known as the WV Partnerships for Teacher Quality (WVPTQ), and operated under this basic structure for the next eight years, with participatory representation and leadership from the groups cited above. For the next several years, the Benedum Foundation and the West Virginia Legislature partnered to continue providing support, building to a budget of nearly $1.1 million by 2009. Since 2009, the initiative has been funded by the WV Legislature.

Fairmont State University, a regional state campus, initiated local partnership work as part of the 2001 efforts. In 2006, this effort moved from a small-scale pilot initiative to a planning process to expand the scope of the partnership with local schools and districts, and to bring teacher education more comprehensively into the collaborative efforts. In late 2007, these efforts led to a strategic plan that expanded the partnership to six school districts (from one) and to 40 local schools (from three). This expansion process occurred with the support of university and district leadership, including the superintendents and school boards. With the expansion, all undergraduate teacher education programs of study were integrated into the partnership structure, with a full range of field and clinical experiences in PDS sites. All local graduate-level teacher education programs (post-baccalaureate and master's-based) were incorporated (only distance online programs were not incorporated). The partnership has benefited from significant new resources, including legislative funding cited above, restructured funding from the university to focus more directly on partnership commitments, and district funds allocated by the local boards of education.

Clearly West Virginia as a state assumed a national lead in building, supporting, and sustaining partnerships to enhance the quality of teacher preparation and professional development. The state, its public schools and school districts, and institutions of higher education made these commitments in the context of limited public coffers and chronic, pervasive poverty throughout the state. As noted earlier, economic barriers and sustained poverty—particularly as they impact children—are central to defining educational challenges in the state. The two partnerships described above, and the state-wide efforts in which they are networked, have moved forward in the face of these challenges.

When one takes into consideration the heightened attention to teacher education, and the more intensive level of intervention in local and state decision making in public education over the last ten years, there is a glaring gap in this success story. There have been no federal funds allocated through teacher quality and partnership grants to support these efforts at either the local or state level. Further glaring light on this gap is created by the intent of federal efforts to highlight support to pockets of poverty through NCLB and federal Title II partnership grant funding. This problem and absence of access to federal grant funding is couched primarily in the limitations in definitions of poverty used by the U.S. Department of Education that are used to identify high-need schools and school districts. (This is compounded by policies, guidelines, and definitions related to Title I.) Because of the limitations of these policies, IHEs are limited in their ability to engage in school-university partnership initiatives that would benefit from Title II grant funding.

Partnerships, Policy, and Poverty

Poverty Policy in Rural Places

As noted by the Rural School and Community Trust (RSCT), poverty, and how it is measured, may not be as reflective of socioeconomic conditions as evidence on the ground in some rural localities might reflect. Evidence cited by the RSCT (October 2009b) suggests that, although poverty in rural places may be more pervasive as a percentage of the population, the statistics may be weighted in such a way that limits rural places' available funding and access to federal resources through Title I. Rural places may actually experience higher rates of poverty, but on a per capita basis receive less funding than urban counterparts, and less than rural communities adjacent to urban centers within a school district. RSCT cites three factors that limit total funds accessed in rural districts. One, variations in Title I funding are due to individual states' differences in wealth and provision of funding through the political process. As the RSCT notes,

> Title I funding varies from state to state based on the average amount each state spends per pupil in its public schools. This is supposed to account for state-to-state differences in the cost of providing schooling. What it really accounts for is state-to-state differences in wealth and in political commitment to education. That disadvantages high poverty rural schools in states that don't spend much on education.
>
> (RSCT, 2009b, p. 1)

Two, the higher the raw number of children in poverty in a district or the higher the percentage, the more a student counts (is weighted) in the process. As a consequence, the more the district receives is a factor of the raw number. Large districts, whether they have a high percentage of children in poverty or not, are advantaged over small districts, even those with high percentages of children in poverty. Three, RSCT references what it terms the "small state minimum." All states, regardless of poverty rates, will receive no less than 0.3% of the total available. The guaranteed base leads to small states with low levels of poverty receiving a disproportionate share of available federal funds.

This effect has been reinforced in the allocation and acquisition of stimulus funds in the most recent economic recession. As with Title I funds, noted by RSCT above, the result is disproportional allocation relative to need across school districts. Smaller raw numbers in small rural districts, even though the percentage of poor students is higher, are allocated relatively less. According to the RSCT:

> That's because the portion of the stimulus funding targeted to the lowest-income students, Title I funding, is distributed through just two of the four Title I formulas. Those two formulas, the Targeted formula and the Education Finance Incentive Grant (EFIG), "weight" a district's student count. That is they provide additional money to districts with larger total *numbers* of poor students.
>
> (RSCT, 2009a, p. 1)

Large urban districts with lower percentage rates of children in poverty, as compared to small rural districts with higher percentages, are allocated more per pupil. To put it in more succinct mathematical terms:

> The poorest 800 rural districts enroll almost a million students and have an average poverty rate of 35.52%. They get a little over $1,200 per poor student in Title I stimulus. Among the seven urban districts with the largest enrollment of poor students, only Detroit has a higher poverty rate than the rural 800 and all seven get more per poor student in Title I funding—in both the regular federal budget and the stimulus package.
>
> (RSCT, 2009a, p. 1)

And the resulting implications, as summarized by the RSCT:

> Per-pupil funding can be problematic for small districts because their relatively small numbers of students mean they often don't accumulate enough funding in any category to cover their costs. In addition, poor rural districts have very few ways to raise revenues locally and they face serious economic challenges that predate, usually by decades, the current recession, so Title I funding is a life line.
>
> (RSCT, 2009a, p. 1)

Teacher Quality Grants: Partnerships Manifested in Policy

The purpose of Teacher Quality grants is to provide funds to support projects that lead to higher quality in new teachers. The two key strategies for this initiative are to improve the preparation of new teachers and to enhance the professional development of practicing teachers. The initiative further supports efforts to lead to greater accountability for the quality of teacher education and to recruit new teachers into the profession (www.ed.gov/programs/tqpartnership/index.html). To support these initiatives, grants may support the creation of school-university partnerships between colleges and universities and school districts that exhibit high need characteristics. "Eligible partnerships" are defined as those that include, among other entities, schools and school districts, and higher education partners from education and the arts and sciences.

As for the definition of high need as related to socioeconomic data, the authorizing legislation (PL 110–115) for Teacher Quality grants uses the phrase "children from low-income families" (www.ed.gov/programs/tqpartnership/legislation.html). The authorizing language defines high need local education agency as the following:

> A High-need Local Education Agency is one where not less than 20 percent of the children served by the agency are children from low income families; That serve not fewer than 10,000 children from low-income families...

These policy implications are reinforced at the state level through Title II funds allocated to states in No Child Left Behind (NCLB) legislation and appropriations.

In West Virginia, NCLB funds provided to support teacher quality within higher education preparation programs are disaggregated to and distributed by the Higher Education Policy Commission (HEPC). The HEPC, led by the Chancellor, provides state-wide leadership for the state's four-year colleges and universities. The HEPC also has policy oversight responsibilities for the institutions of higher education (http://wvhepcnew.wvnet.edu). The HEPC provides agency logistical support for NCLB funds allocated to the state, with the specific purpose of supporting higher education policy implications of the federal legislation.

For the 2010–2011 fiscal year, the HEPC has allocated, under the federal guidelines, $578,399 for its "Improving Teacher Quality" State Grants Program. Colleges and universities may submit proposals in response to requests to access these funds to support strategies to enhance the quality of teaching and teacher preparation. To be eligible for funding, proposals must include evidence of the following partners:

- A college or university-based educator preparation program;
- A college or university-based arts and sciences program;
- A "high need" school district.

Consistent with federal policy, a high need school district is defined as "serves not fewer than 10,000 children from families with incomes below the poverty line" or "For which not less than 20 percent of the children served by the agency are from families with incomes below the poverty line" (www.wvhepc.org/academic/RFP_2010_Packet.pdf).

These conditions are coupled with requirements related to the percentage of teachers not teaching in the field for which they were prepared, and the percentage of teachers on emergency permit or licensure (www.wvhepc.org/academic/RFP_2010_Packet.pdf). As the guidelines point out, given these requirements, only 24 of the state's 55 county school districts qualify as a potential partnering school district. (The guidelines do stipulate that if one such district is a partner, then districts that do not meet the guidelines may be included in the project.)

Following are some of the implications in the implementation of this policy translated into funding guidelines for school-university partnerships and school districts in the state:

- Of the 10 counties that are home to a public college or university (those described in the state-wide partnership discussed above and that prepare the vast majority of the state's new teachers), 6 are not eligible for grant funds.
- Of the 24 eligible counties, 13 do not share a border with a county that is home to a public college or university.
- Of the 24 eligible counties, 12 are clustered in one region of the state (the lower third geographically), in an area with the greatest geographical challenges, and with no college or university in a central location in the shared area.

- Of the 35 non-qualifying counties, 17 are either home to a public college or university, or border a county that is.

In summary terms, the more economically challenged an applicant may be, and the more educational institutions could benefit from a school-university partnership supported by federal grants under Title II, the less likely they are to be eligible for such a grant opportunity. In addition to the stipulations cited above, these "high need" districts (and any others that participate in a project) must also be able to put on the table a real-dollar commitment (beyond any in-kind commitments). Finally, as part of the partnership agreement, no specific partner (high need or otherwise, K-12 or higher education) may have at their specific disposal any more than 50% of the grant total.

For teacher education based in higher education, where partnerships are central to the education of educators, contextual factors related to resources are critical. Schools, school districts, and institutions of higher education (IHEs), in many cases with partnership-based educator preparation and professional development—certainly those in West Virginia—have by design built programs immersed in and engaged with these local contextual factors. A foundational aspect of these programs is to contextualize (1) the conditions in which educators practice into the preparation/professional development programs and (2) the process of renewal and creating best practices. For areas that struggle with chronic poverty and resource limitations, external resources play a vital role in creating new structures and relationships for creating better schools, preparation programs, and excellence in educator practice.

For many alternative providers, particularly those that would be possible under Race to the Top provisions, contextual factors are minimal, if relevant at all. For online and market-based providers in particular, workshops, professional development, and course content may be so decontextualized as to be offered on a national scale, with no effort to engage local context, and the challenges of embedding practice, professional reflection, and professional development in the realities of teaching and schools. For such programs, building a resource base to be innovative, and securing resources to be creative in a challenging context, are irrelevant to providing a professional service. What can be delivered online in rural Appalachia can be delivered online in urban Houston, and with the potential economy of scale at "rock-bottom prices" as sales pitches go.

It is in this set of challenges where federal policy around Title I and II could be most helpful, but because of a one-size-fits-all definition of poverty and how it gets manifested at the local level, resources do not make it to that local level and to the most contextualized, innovative work. Spaces of innovation and creativity require different investments of capital as any business would know, but in the transfer of business practices to public education, this tenet does not carry over. Innovation, creativity, and efforts to meet the needs of a local market are sacrificed in the name of cost-effectiveness, policy efficiency, and mile-wide, inch-deep reform. Of possibly even greater importance, the creation and sustainability of institutional relationships, and teachers and schools grounded in local needs and opportunities in many cases, is marginalized. In such cases, where new resources could have been offered or acquired through partnerships, many children get left behind because of a presumption of adequate resources (as defined in policy terms). And, there are too few opportunities for them to acquire better new teachers, and to support the ones they have in building on their knowledge and skills in ways that fit local and community needs.

It is also important to note that such efforts on the part of rural IHEs and schools to build partnerships with limited resources are further complicated by the demands of state and national accreditation. This is exacerbated greatly for more intensively rural and resource-strapped school districts such as many in West Virginia, and the IHE partners, who choose to work in school-university partnerships. These partnerships and their educator programs are complex entities, requiring new structural and organizational arrangements, and relationships and initiatives that are more labor intensive and sophisticated than traditional go-it-alone strategies. These structures, relationships, and strategies must and should be built into processes such as NCATE and TEAC accreditation. Such processes must be sensitive to those sometimes profound differences, and to the resource challenges that are created in resource-poor areas. For programs not based in these partnerships, such matters are not an issue. For programs based in school-university partnerships, such efforts—with limited resources—may not be acknowledged.

Federal policy on ESEA Title I and II, and in how such funds are allocated, may and likely do alter the educational and professional paths for those who are the children in rural places, and who chose to be the teachers for those same places. In rural places, many if not most teachers call the local area (if not the specific community) home. They grow up in many cases in resource-limited schools and districts with fewer supports and advantages than their counterparts in other districts with greater and more diversified funding. They often attend a regional college or university for their teacher education, and the IHEs themselves are limited in their resources as a result of a conservative tax and funding base in the state. And, although West Virginia regularly ranks high in expenditures on public education as a factor of available resources, many schools, districts, and IHEs must operate on a conservative fiscal margin. They then graduate from their college or university as a teacher to work in school districts that have limited resources to support professional development and continued professional learning. If such schoolchildren, teacher education candidates, and practicing educators do have the advantage of being in an area where a school-university partnership is a potential resource, as noted above, that partnership may face significant challenges in securing critical Title II grant resources to support the work. This is a double whammy, as these schools and districts will face parallel limitations in securing Title I resources as well.

From "One Size Fits All" to "Tailored Design": Making Policy Development and Implementation Work in Context

A vital element of teacher preparation and the support of practicing teachers is to get to elements of practice—the knowledge, skills and dispositions one needs to be able to teach well—and to *apply and adapt those skills to contextualizing factors*, such as: urban, rural, suburban; racially/ethnically diverse; high poverty, resource poor. This premise intersects all contexts, and most certainly is shared across the most challenging rural and urban communities. No doubt differences between the two are in many cases profound, and in some instances subtle, but the ability to context- ualize preparation and practice to meet the needs of children and to support educators in doing so is consistently essential to educator preparation and pro- fessional practice. All issues related to demand for teachers, availability of teachers, who and how they are prepared must rest on this basic contextual assumption if we are to ensure that every child has access to a creative and talented teacher, regardless of who they are, where they live, and the conditions of their lives.

This understanding of the significance of contextualization of practice provides a strategic intersection for rural and urban coalitions and for boundary-spanning for educators and policymakers. What both broader communities need is an element in the construction of federal policy and grant-based opportunities that allow for (if not promote) variation in the definition of the local challenges and the parallel application of innovation and creative initiatives. By moving away from "one-size-fits-all" assumptions, and allowing for this variation, federal policy may actually be able to support and make real an assumption that pathways to success can cross all communities, but will take different routes in rural and urban contexts to get to success. Success may actually in the end share essential elements, with contextual streets, avenues, and country roads to get there. Such flexibility may also give and promote greater flexibility at the state level in the application of federal policy, and the use of federal funds, in supporting local needs, local innovation, local creativity, and local success. Such could also be true for national organizations, particularly those related to accreditation and support of prepara- tion programs and professional interest groups. These agencies and organizations may feel less pressure to compromise their missions and agendas to meet the demands of limiting federal policies, and create a greater sense of opportunity and risk-taking to get to better practice and opportunities for children and educators.

No doubt efforts through federal teacher education policy such as those asso- ciated with the implementation of Title I and Title II, meeting the demand of new teachers in areas with critical shortages, and even Race to the Top, are intended to help educators meet challenges and barriers in the education and professional development of educators. Even with the current policy structures, and given the limitations described above, such resources can and do have a posi- tive impact in local and episodic cases. But, as noted in the context of West Virginia as described above, and Appalachia more generally, the broad sweep that federal policy may be designed to achieve (providing resources to schools, districts and IHEs with concentrations of poverty) gets lost in the translation to local demographics and the communities in which children, adults, and families live. And, as in the case of New York, efforts to change the dynamic of how we prepare and ensure enough teachers for communities without access to them may limit the range of possibilities for how institutions of higher education can be a resource to solving that problem.

As noted previously, federal policy limitations can drive parallel limitations at the local and state level. In West Virginia, efforts to support educator professional development and preparation, and initiatives to support school–university partnerships may be hampered in access to, if not excluded from, garnering resources related to Title II funding. As in New York, pathways to teaching that presume that all that matters is expediency and the "reality of the classroom" may disenfranchise institutions with outstanding histories of educating educators. Although likely not intended to be monolithic in its impact, such policy constraints are monolithic in their application. Poverty is defined as a standard in essence, not as a contextualized factor in the lives of children and the work of educators. Access to teachers becomes defined as a near-open-door policy with increasingly limited safeguards in public trust as to who can become a teacher. The resulting implication is that state departments of education and governing bodies for higher education (as is the case in West Virginia) translate federal pressures created by policy definitions and limitations down to institutions in "one-size-fits-all" strategies to be in compliance with federal mandates guidelines. In this picture, efforts by local and regional institutions to address local and regional challenges sit in the crosshairs of decontextualized crossfire. Their constituent partners are expected to meet the policy mandates, while at the same time they are marginalized in their ability to access resources to engage in the work in a creative and innovative way, or to potentially not be at the table at all (as in New York).

The case of New York is particularly illuminating on this point about constituent partners meeting a local need. To repeat a position held by the Board of Regents in Policy §237(2), part of the purpose of policy related to higher education is to "describe statewide needs, problems, societal conditions and interests of the citizens and discuss their priorities." One of the planning goals for higher education in New York is to address the state's challenges related to quality of life for the state's citizens. To do so, programs need to be grounded in a rich context engaged with those challenges. Outside providers of any academic programming, in particular in professional fields, would face greater challenges in that engagement. As a policy issue—with implementation, public trust, and the public good as central framing—these constructs are inherently local, and require an intersection between the broad nature of state and federal policy and the ability to make it work in implementation at the local/regional level. A critical policy question is how providers from outside the state, who design programs for broad national market appeal, can tailor programs to meeting local needs or at a minimum contextualize these constructs in business models? If providers outside New York can meet this condition, it may raise a question among policy critics: Doesn't this render the Regents on this matter irrelevant? What we really need (nationally or federally) is a national board that can vet national providers. Or, as a reverse question: If national providers are simply mirroring what local/regional/state institutions can do, why go to national providers in the first place?

A policy critic might also raise this question. The Regents plan for approving new programs sets a precedent for what would become the emergency plan for alternative licensure, but the process would also have allowed for a review process to expand certification through existing institutions. The emergency authority to approve new programs outside higher education is out of sync with the historical policy. If the process has worked to ensure quality, then stepping outside it would jeopardize the integrity of programs in general, and certainly that of a program approved in this way. If that were not true, how could the public have faith in the process to assume that any program would have integrity, particularly those that step completely outside of structures created to protect public interest? Authorizing outside providers would fly in the face of the social, contextual factors that are foundational to the program approval process in New York. The option for endorsing outside providers runs counter to the more general program approval process, and decontextualizes the policy implementation. Outside providers from outside New York do not have to contend with histories, community needs, or localized issues. In essence, do the needs of all the people of New York get disengaged from, if not lost altogether, in the effort to create the perception of meeting the needs of a perceived market, or defining that market (real or imagined) in the absence of how state/federal policy gets interpreted at the local or regional level within the state?

The focus on the contextual factors and significance of policy implementation, as opposed to design, are paramount. However, these points should not be interpreted to singularly privilege local context and perspective, or to romanticize it as infallible. Clearly the history of major federal legislation and policy implementation is fraught with misinterpretation, misapplication, and intentionally mischievous behavior. Local policy implementation—and its manifestation in institutional missions and work—left to their own devices do not, in and of themselves, lead to good results for children, educators, schools, districts, and communities in the absolute. The point is that it is impossible for policy (federal, state, or local) to work in a meaningful way if it is not connected to the local context in which meaning is made and implementation has to occur (i.e., in the application of the policy). It is equally problematic if the definitions on which the policy is built (e.g., poverty or access to enough good teachers) is defined out of the on-the-ground contexts that clearly are what the policy is meant to support.

Policy should be consistent in identifying the problem to be solved, the opportunity to be created, and the intended outcomes. It should be variable in its flexibility for how the problem gets manifested locally, what the opportunities (given the resources the policy presents) to be created are, and how generally intended outcomes might get evidenced in artifacts at the local level. Federal policies create pressures that tend to decontextualize practice, and we need to construct in the policy process elements that allow for contextualization of practice to get to a set of definitions of success and opportunity on which we likely have consensus. It should also create space, for example, in the context of both policy contexts described, for:

- Reversing the trend toward national policy solutions that marginalize local context to a more balanced immersion of policy definition, development, and implementation in local need and context.
- Ensuring that issues related to teacher quality, access to excellent teachers for all learners, professional support, and mutually beneficial partnerships between P-12 schools and higher education are framed in ways that fit the need rather than fit the bureaucratic (and sometimes political) need to standardize policy;
- Defining policy challenges and solutions to problems in local terms, and in ways that can be tailored to solving local problems and creating local opportunities;
- Ensuring that local and regional partners have adequate resources to educate local children, who are likely to go to local colleges, and to return home to be teachers in the local area.

These two portraits—one rural and one urban, and in juxtaposition to each other—help to provide important insights into how we as a country can do a better job of creating and implementing constructive federal policy that impacts teacher education specifically. These insights may also be helpful in the creation of general education policy. They may also help us to better understand how we can do a better job of facilitating the interplay of federal policy with state and local needs, and a better job of tailoring federal policy to the local implementation challenges and opportunities. As a resource issue alone, federal policy has been, and likely always will be, an important catalyst for important school renewal initiatives at the state and local level. The challenge is how to balance the need for resources, and the federal government's ability to provide them, with local and regional variation in a context that will be significant to the success of any policy action. The more economically challenged a local or regional context might be, the more dependent educators at the P-12 and higher education levels will be on those resources to generate meaningful change, or to sustain best practices. A central question, given the two very different contextual challenges of urban and rural America is: Does one policy (on any given issue or challenge) work for both? The answer is yes, if the implementation process allows for enough variability given the local context for implementation, and the flexibility to engage the partners, players, stakeholders, and broader community members who should have a vested interest in the policy's promise, and the impact it can have on their children, their children's educators, and their local communities. If that tailoring can lead to that success, then the effort, translated to the state and federal level, stands a much greater chance of success.

References

American Association of Colleges for Teacher Education. (2010). *The clinical preparation of teachers: A policy brief.* Washington, DC: AACTE.

Annie E. Casey Foundation. (2010). Kids count data book: State profiles of child well-being. Baltimore, MD. Retrieved August 20, 2010 from www.aecf.or/media

Cross, C.T. (2004). *Political education: National policy comes of age.* New York: Teachers College Press.

DeHaan, L. & Deal, J. (2001). Effects of economic hardship on rural children and adolescents. In R.M. Moore III (Ed.), *The hidden America: Social problems in rural America for the twenty first century* (pp. 42–56). London: Associated University Presses, Inc.

Dempsey, V. (2007). Intersections on the back road: Class, culture and education in rural and Appalachian places. In J. Van Galen and G. Noblit (Eds.), *Late to class: Social class and schooling in the new economy* (pp. 287–312). Albany, NY: State University of New York Press.

Education Week. (2010). Quality counts: West Virginia state highlights. Retrieved from www.edweek.org/products

Foderaro, L.W. (2010, April 20). Regents plan new route to master's in teaching. *New York Times.* Retrieved from www.nytimes.com

Fraser, J.W. (2007). *Preparing America's teachers: A history.* New York: Teachers College Press.

Michelli, N. (2005). The politics of teacher education: Lessons from New York City. *Journal of Teacher Education 56*(3), 235–241.

New York State Education Department/University of the State of New York. (2010). April 19, 2010. Albany, NY. Retrieved from www.regents.nysed.gov/meetings/2010Meetings/April2010/0410hea4.pdf

Noblit, G. & Dempsey, V. (1996). *The social construction of virtue: The moral life of schools.* Albany, NY: State University of New York Press.

EDITORS' COMMENTARY

One cannot help but be struck by the descriptions of the limitations of federal policy and perhaps the unintended outcomes of such policy as they affect those children with the greatest needs: those in urban centers like New York City and those in rural settings such as described in West Virginia. Decisions in pursuit of Race to the Top funding in one case, and the definitions of eligibility for funding embedded in policy in another, seem to these two deans not to take into account the real needs of these children or the local needs of their contexts.

New York State higher education institutions responded to the 1999 policy changes in certification—creating what some have seen as the most stringent in the country—by resubmitting every program and pursuing national accreditation. For many institutions, national accreditation (which included the options of accreditation by NCATE, by TEAC, and by a Regents-created Regents Accreditation of Teacher Education system) was their first effort to seek accreditation. Only a handful of New York institutions were in fact accredited at the outset. All were accredited by 2005. Teacher educators in New York have called for an assessment of the impact of the new policies, while colleges and universities developed their own systems of evidence in pursuit of accreditation. With the requirements of Race to the Top funding under consideration, the Regents appeared to have undermined their own system first by empowering not-for-profit non-collegiate agencies which seem unlikely to be able meet the standards to offer certification set by the Regents themselves and, second, by moving to use the very narrow measures of value-added assessment as the coin of the realm in the assessment of teacher education. Both of these policies were dictated by the pursuit of Race to the Top, and that pursuit was successful, yielding some $700 million. Local perspectives on why we educate, developed by colleges in collaboration with school districts as described in the case of Brooklyn College, seem not to matter. It also seems that compliance with all of the expectations of the state including accreditation were not expected. Responding to letters of inquiry and protest from a number of New York institutions, the Regents and the Department appeared prepared to slow down the process.

A similar lack of consideration of local circumstances seems to have prevailed in West Virginia. The federal definitions of poverty enabling IHEs to be eligible for grants to pursue partnerships and meet the needs of the schools they served made that very collaboration impossible.

We would hope that these were unintended outcomes. More cynical commentators might argue that children and families in poverty play a lesser role in politics, allowing for their needs to be ignored.

Another element of the West Virginia case contrasts the ability of the state's flagship university to seek and secure external funding to develop a strong, nationally recognized partnership, an option not available to the smaller institutions serving rural populations, because access to funding are so limited.

Discussion Questions

1. What steps might be taken to assure that local needs are met in establishing guidelines for federal policy involving the distribution of resources? Would allowing institutions to make the case for exceptions to rules for awards violate the fairness of access to funds, or would it make the access fairer?
2. Who can or should provide oversight to rulemaking when there are seeming violations of equity at the state or federal levels?
3. From what you see here and in Chapter 7, how are the Benedum partnership and the Montclair partnership similar and different?
4. What is the impact on professionals and pathways when rules are changed or

ignored after institutions have labored successfully to meet complex high standards?

5. Do these cases illustrate a lack of concern, disdain, or even contempt for teacher education programs that link colleges and schools whether in urban or rural areas? Is there some other more plausible explanation for the actions described?

9

TEACHER EDUCATION POLICY AND SOCIAL JUSTICE

Marilyn Cochran-Smith and Kim Fries

This chapter is about teacher education policy and social justice. Accordingly, we begin with a brief discussion of our perspectives on each of these. Then we take up three examples of contemporary teacher education policy that are related to issues of social justice. These three examples differ from each other in terms of the contextual levels at which they occur, the agents or agencies that are central, the larger professional and political agendas that are involved, and the problems, challenges, and paradoxes raised about social justice. Across the three examples, we suggest that although there are local pockets of institutional policy focused on social justice, the current stance of national accreditors is unenthusiastic at best, and the orientation underlying state and federal policies is based on a narrow conception of justice as redistribution of teacher quality and other school resources. This orientation does not account for the larger historical, systemic, and structural issues that maintain educational injustice and inequity, nor does it acknowledge the capacity of teachers to be advocates, agents of social change, and collaborators with others in larger movements for social justice.

A Perspective on Teacher Education for Social Justice

Although social justice is a common theme in many teacher preparation programs, there is considerable variation and ambiguity about the meaning of the term. Despite a lack of conceptual clarity in the field, in this chapter, we work from Cochran-Smith's (2010) theoretical framework, which conceptualizes teacher education for social justice as a coherent and intellectual approach to the preparation of teachers that acknowledges the complex social and political contexts in which this is located. Cochran-Smith argues that a theory of teacher education for social justice must address three critical interdependent questions: What do we mean by justice? How do we think about teaching and learning in a way that enhances justice? How do we conceptualize and assess teacher education that prepares teachers to foster justice and supports them as they try to live out this commitment in school settings?

To do so, a theory of teacher education for social justice is made up of three overlapping subtheories: (1) a theory of justice that makes explicit its ultimate goals and considers the relationships of competing conceptions of justice; (2) a theory of practice that characterizes the relationships of teaching and learning, the nature of teachers' work, and the knowledge, strategies, and values that inform teachers' efforts for social justice; and (3) a theory of teacher preparation that focuses on how teachers learn to teach for justice, the structures that support their learning over time, and the outcomes that are appropriate for preparation programs with social justice goals, including students' learning outcomes. We briefly take up each of these below.

A Theory of Justice

A distributive notion of justice dominated theories of social justice for the latter half of the last century (Fraser, 2003; Howe, 1997). However, the "politics of difference," which emphasized recognition rather than redistribution (Young, 1990), made it clear that failure to recognize and respect social groups was a central dimension of injustice, and thus the goal of recognition had to be central to justice theories. In contemporary political philosophy, the critical question is how to conceptualize the relationship between the notion of distributive justice that is central to modern liberal democracies in relation to the politics of identity and difference (e.g., Fraser, 2003; Gewirtz & Cribb, 2002; Honneth, 2003; Young, 1990). Cochran-Smith's effort to answer this question for teacher education is informed by Howe's (1997) political understanding of what equality of educational opportunity requires; Reich's (2002) integration of liberalism and multiculturalism; and King's (2008) argument for drawing on the knowledge traditions and lived experiences of marginalized and oppressed groups.

The notion of justice in Cochran-Smith's framework for teacher education emphasizes equity of learning opportunities and outcomes for all students with simultaneous challenging of school and societal practices that reinforce inequities; recognition and respect for all social groups with integration in classrooms and schools of the knowledge traditions and ways of knowing of marginalized groups; and direct acknowledgment of the tensions and contradictions that emerge from competing ideas about the nature of justice and efforts to manage those tensions in actual educational settings.

A Theory of Practice: Teaching and Learning Relationships

The second question in Cochran-Smith's theory of teacher education for social justice is: How can we conceptualize teaching and learning practice in a way that enhances justice? The answer to this question is essential because it is the inter-mediate link that connects teacher preparation and justice. In other words, a theory of teacher education for social justice must have a well-theorized idea about the kind of teaching practice that enhances justice since preparing teachers for practice is the goal of all preparation programs and pathways.

Cochran-Smith's argument is that, in order to support justice, teaching practice must be theorized as an amalgam of knowledge; interpretive frameworks; teaching strategies, methods, and skills; and, advocacy with and for students, parents, colleagues, and communities—all with the larger goal of improving students' learning and enhancing their life chances. The idea of practice as defined by teachers' knowledge, interpretations, methods and advocacy emphasizes that practice is theoretical as well as practical, critical as well as relational.

A Theory of Teacher Education: Teacher Learning Over Time

The third question that must be answered by a theory of teacher education for social justice is: How can we conceptualize teacher preparation intended to prepare teachers to engage in practice that enhances justice? Again, the answer to this question is central because it reflects the direct link between teacher prepara-tion and teaching practice. Cochran-Smith's argument is that, in order to support teaching and learning practices that foster justice, teacher preparation must be theorized in terms of four key issues: who should teach, which is instantiated in practices and policies related to the selection and recruitment of teacher candi-dates; what teachers and students should learn, which plays out in the curriculum and pedagogy to which teacher candidates are exposed; how and from/with whom teachers learn, which has to do with the intellectual, social, and organiza-tional contexts and structures designed to support candidates' learning; and how all of this is assessed, or how the outcomes of preparation are constructed and measured and what consequences these have for whom.

The overarching idea in theorizing teacher preparation for social justice is that it is intended to challenge the educational status quo and be transformative. In sharp contrast to preparation intended to be ever more closely aligned with the accountability system, teacher preparation for social justice challenges the testing regime and the inequities it reinforces. This does not mean that teacher education for justice simply sits outside the accountability system, however. Rather the point is to construct a different kind of accountability by working simultaneously within and against the system.

A Perspective on Teacher Education Policy

In this chapter and elsewhere, we work from a discourse approach to the analysis of teacher education policy (Cochran-Smith & Fries, 2001, 2005, 2011). Since the 1990s, a discourse approach to policy analysis has been used in social science fields, including political science, sociology, linguistics, planning and environmental policy, nursing, and education (e.g., Bacchi, 2000; Ball, 2008; Cheek & Gibson, 1997; Fischer & Forester, 1993; Joshee, 2007; Luke, 2002; Popkewitz & Lindblad, 2000; Sharp & Richardson, 2001).

A Discourse Approach to Policy

Generally speaking, those who take a discourse perspective reject the idea that policymaking is the result of the objective and non-biased assessment by experts about how to obtain clear and fixed goals; they also reject the idea that policymaking is an apolitical, strictly rational process. Rather, from a discourse perspective, it is understood that goals themselves are competing and protean. Problems are constructed by multiple actors rather than discovered "out there" through universal and scientific methods. From a discourse perspective, it is assumed that policymaking involves many agents at multiple formal and informal levels, all of whom are engaged in constructing meaning (Bacchi, 2000). The result, as Deborah Stone (1997) suggests, is that policymaking is always a struggle over ideas: "Each idea is an argument ... in favor of different ways of seeing the world ... There are multiple understandings of what appears to be a single concept, how these understandings are created, and how they are manipulated as part of political strategy" (p. 11).

The Argumentative Turn

Echoing the language of the "interpretive" and "linguistic" turns in 20th-Century philosophy, Fischer and Forester (1993) use the term "the argumentative turn" to emphasize that policymaking and policy analysis are argument-making processes. Like Stone, they emphasize that policy actors must first formulate and construct what "the problem" is before they can propose plausible solutions. In political terms, they suggest:

> policy and planning arguments are intimately involved with relations of power and the exercise of power, including the concerns of some and excluding others, distributing responsibility as well as causality, imputing praise and blame as well as efficacy, and employing particular political strategies of problem framing and not others.
>
> (1993, p. 7)

A central way groups, individuals, and agencies promote their definitions of problems and their conclusions about who is praise- and blame-worthy is through metaphor and analogy, emblematic language, symbols, stories, and literary devices along with recurring arguments that forward their own positions and discredit others. All of these can be understood, discursively, as attempts by the proponents of particular positions to garner support—not simply for the solutions they favor but also for their ways of understanding the issues in the first place (Stone, 1997). Viewing policy through a discursive lens acknowledges that policy controversies are struggles over values, world views, and underlying ideologies as well as agreements and disagreements over strategies.

The Teacher Quality/Teacher Education Policy Web

Consistent with a discourse perspective on policy, in our consideration of teacher education policy we use the notion of a "policy web," as developed by Reva Joshee and Laurie Johnson (Joshee, 2007, 2009; Joshee & Johnson, 2005), who in turn built on Hogwood and Gunn's (1990) suggestion that policies are made "by the interactions of many policy influentials operating in a power network" (as cited in Joshee & Johnson, 2005, p. 55). Joshee and Johnson use the web image to convey the idea that policies are developed formally and informally on multiple levels and in multiple forms, like the rings of a web, and that policy discourses are both discrete and interconnected, like the cross-cutting, but non-linear, threads of a web. The web image also suggests that specific policy issues must be interpreted within a larger network of related policies. As Joshee and Johnson point out, a policy web calls attention to the relationships between and among discourses, who the actors are, how new ideas and competing agendas enter into the larger arena, and which discourses are predominant, silenced, valorized, and marginalized.

Increasingly over the last decade, major policy discussions about the preparation of teachers have been part of, and/or intertwined with, larger discussions about teacher quality. In fact, at this juncture, it is impossible to understand teacher education policy without understanding larger teacher quality issues. Like policy regarding other public services in the U.S., policies related to teacher quality and teacher education are not developed and enacted at a single level by a single agency, but at multiple levels and by many actors, including federal, state, and local agencies. In addition, teacher quality/teacher education policy is developed and enacted by professional organizations and national/regional accreditors as well as by individual higher education institutions (or higher education systems) and by alternate providers of preservice preparation who make decisions related to recruitment, admissions, placement, curriculum, program completion, and graduation.

Teacher Education Policy and Social Justice: Three Contemporary Examples

In keeping with the viewpoint that teacher education policy is enacted at multiple levels and by multiple agencies, the three policy examples we examine are drawn from different levels—(1) the level of federal/state education policy regarding teacher quality and teacher education, (2) the level of standards and policies established by national teacher education program accreditors, and (3) the level of local policies and practices developed by individual universities or other agencies that prepare teachers. We use these examples to explore critical issues related to teacher education policy and social justice and to demonstrate that policy issues must be interpreted within a larger network of related policies.

Example 1: Federal and State Policy Regarding Teacher Education

Over the past decade, the recruitment, preparation, and retention of teachers has become a central concern of top-level leaders in the business, philanthropic, and policymaking worlds as well as in the professional arena in the U.S. and elsewhere (Cochran-Smith & Fries, 2005; Oakes, Lipton, & Renee, 2006). Along these lines, there has been a steady stream of position papers, think-tank reports, and new policies or policy proposals regarding teacher quality and teacher preparation along with a great deal of controversy about these. To sort this out, we conducted a discourse analysis of more than 225 formal and informal policy documents and other materials produced between 2005 and 2010, including news articles, blogs, op-ed pieces, podcasts, Congressional testimony, public position statements, and reports regarding U.S. teacher quality and teacher education (Cochran-Smith & Fries, 2011). Our first example of teacher education policy and social justice draws from this analysis.

Not surprisingly, our analysis revealed that there is not a single policy discourse about teacher education and teacher quality at the state and federal level in the U.S., but multiple discourses, which compete with one another, are also sometimes combined. We identified five major discourses, which we labeled: (1) the teacher quality gap and educational inequality, (2) teacher quality and the market, (3) teacher quality in the knowledge economy, (4) teacher quality and professional teacher education, and (5) teacher quality and social justice. The argumentative structure of each of these is based on a particular construction of the issues regarding teacher education and quality, which leads to particular policy recommendations. These discourses reflect larger world views and ideologies related to justice, equity, and diversity

We concluded that the first and third discourses, which are often braided together in policy debates to achieve powerful rhetorical force, have become dominant over the last decade and are far-reaching in influence. Meanwhile the second discourse has become somewhat less influential than previously, although it is still prevalent, and the fourth and fifth discourses have been marginalized. Below we elaborate on the two dominant discourses regarding teacher quality and teacher preparation and then consider these in terms of social justice issues.

The Teacher Quality Gap and Educational Inequality

In policy debates about teacher quality and preparation, the "teacher quality gap" has emerged as a powerful idea that builds on the connotations and language of the "achievement gap." The argumentation of this discourse goes something like this: (1) Research has shown that teacher quality and effectiveness are among the most important factors in students' achievement. (2) Schools with large numbers of poor and minority students are most likely to have teachers who are not well qualified. (3) Thus the teacher quality gap exacerbates the achievement gap. (4) Direct action targeted at school factors is needed to redress the inequality of educational opportunities and outcomes, particularly distribution of quality teachers.

Underlying the *teacher quality gap* discourse is the assumption that equal opportunity is part of equality, and the opportunity to be taught by well-educated teachers, which will presumably lead to equal educational outcomes, has too long been denied to students in hard-to-staff and low-performing schools. This is dramatically different from simply claiming that all students must achieve to the same high standards despite unequal opportunities. This discourse is illustrated in a number of widely disseminated reports and analyses about teacher distribution patterns and states' responses to No Child Left Behind's (NCLB) equity requirements.

From the perspective of the teacher quality gap discourse, blame for inequalities goes to federal, state, and school district policies—especially teacher licensing policies and teachers' union contracts (Hess, Rotherham, & Walsh, 2004)—and other "anti-performance structures" that preserve a "failed system" (Education Equality Project, 2008, 2009). Here, university-based preparation programs are generally regarded as barriers to equality, in part because they "make excuses" for teachers' failure to close achievement gaps (Haycock, 2004). The argument here is for replacing the status quo with what are referred to as "progress-based" rather than "inputs-based" approaches (Education Equality Project, 2008, 2009), such as: alternate entry routes into teaching; new data systems tracking teachers' effectiveness, students' achievement scores, and teacher preparation; revised hiring and assignment practices; rewards connected to effectiveness; and improved mentoring.

Underlying the teacher quality gap discourse is a distributive notion of justice (Fraser & Honneth, 2003; North, 2006), wherein teacher quality and effective schools are regarded as goods and services that have hitherto been unequally distributed in society. However, there is little acknowledgment in this discourse that educational inequality is rooted in and sustained by larger societal inequalities (Fraser, 2003), including unequal access to healthcare, early childhood services, housing and transportation, and job development initiatives. There is also little recognition that educational goals and curricula fail to reflect the values and knowledge traditions of marginalized groups. Rather, a premise of this discourse is that the remedy for inequality is ensuring that everybody has access to the existing system, more or less assuming that those who are currently "unequal" want to be like the dominant group and will be once they have equal access to teacher quality and effective schools. Ultimately, this discourse represents an ideology of liberal democracy (Joshee, 2007) with an underlying view of diversity as something to be overcome or neutralized.

Teacher Quality in the Knowledge Economy

The discourse we refer to as "teacher quality in the knowledge economy" is prominent in many state and federal discussions about teacher quality and preparation and is sometimes present in the language and logic of national professional organizations, accreditors, and other teacher education leaders as well. Its argumentation is as follows: (1) We live in a globalized society with a knowledge economy that requires world-class academic standards and cognitively complex skills for problem solving and decision making. (2) The robustness of the economy depends on the country's educational achievements, which in turn depend on the quality of its teachers and schools. (3) However, both international comparisons and national assessments indicate that teachers are not teaching to world-class standards and that large segments of the school population are not prepared for work or higher education, which costs the country in individual productivity and economic growth. (4) Radical changes are needed, including rigorous new standards and assessments for all students, a more talented and effective teaching force, and a drastically revamped system of continuous, evidence-based teacher education.

This discourse, which links quality of teaching and educational achievement, on one hand, and the values of progress, global leadership, and economic prosperity for all, on the other, has become a litany in policy debates since *A Nation at Risk* was released in 1983 (National Commission on Excellence in Education, 1983). This same logic is the rationale for the education monies in the American Recovery and Reinvestment Act (2009) and is replete in statements by President Obama, Secretary of Education Arne Duncan, and other administration officials. This discourse is also reflected in the Common Core State Standards Initiative, a project to adopt rigorous common standards for high school students across the states (National Governors Association, Council of Chief State School Officers, & Achieve, 2008).

The knowledge economy discourse uses a "story of decline" (Stone, 1997, 2002) to construct the educational status quo (including "business as usual" at schools of education) as both the cause of the current deterioration of the country's international standing and the major obstacle to change (e.g., Duncan, 2009c; National Center on Education and the Economy, 2006; Vagelos, 2006). From the perspective of this discourse, specific solutions to the problem of teacher quality and preparation include: systematic state-level data systems that link student, teacher effectiveness, and teacher preparation data; recruiting more talented teachers, especially through alternate pathways; training teachers to use data for continuous improvement; national assessment of teacher candidates; accreditation standards dependent on student outcomes; explicit teacher training for effective practice; and policy decisions about preparation programs and pathways based on evidence about outcomes (e.g., Cochran-Smith & Zeichner, 2005; Duncan, 2009a, 2009b, 2009c, 2009d; National Academy of Education, 2009; Wineburg, 2006). All of these proposed solutions to the teacher quality problem zero in on outcomes and accountability for students' learning.

Some of the new outcomes and accountability emphases of professional organizations such as the National Council for Accreditation of Teacher Education (NCATE) and the American Association of Colleges for Teacher Education (AACTE), which have long been regarded as "the establishment" of teacher education, also reflect the knowledge economy discourse. These new directions may be understood, at least in part, as efforts to align with the powerful knowledge economy discourse, which dominates many federal and state policy debates about education and the now firmly entrenched system of accountability that is driving it. Although it is likely that leaders of these professional organizations are genuinely committed to improving student learning outcomes, their recent efforts may also be seen as what Penelope Earley (2000) once described as avoiding being "cast as a culprit" in the larger phenomenon of lower scores on international competitions.

Federal/State Policy Discourse and Social Justice

The teacher quality gap discourse and the teacher quality in the knowledge economy discourse have become dominant in state and federal policy discussions about teacher quality and teacher preparation. Increasingly these two are also braided together, which is not surprising given the influential role the business community plays in current education policy. For example, the influential McKinsey report (McKinsey & Company, 2009) identified four distinct gaps in U.S. education (i.e., between the U.S. and other countries, white students and black and Latino students, students from different socioeconomic groups, and students attending different school systems). The McKinsey report characterized these gaps, not simply in terms of civil rights or the requirement of a democratic society, but as "the economic equivalent of a permanent national recession," with serious consequences in earnings, health, and incarceration rates for individuals. Likewise, in a report jointly sponsored by the National Governors Association, Council of Chief State School Officers, and Achieve (2008), the business agenda and the rights agenda were explicitly joined: "Recent studies offer compelling evidence that educational equity is just as important for economic competitiveness as it is for social justice" (p. 14).

As these quotations suggest, when the two dominant discourses are linked, the bottom-line message about teacher quality policy is clear: Access to high quality teachers must be redistributed in the U.S. so that every child and adolescent, including the nation's increasingly diverse population and segments of that population that historically have not been well served by the educational system, has equal opportunity for a good education. The assumption is that redistribution of access to teacher quality, which will also lead to equal outcomes, will result in everybody being prepared for work and/or higher education, and thus everybody being able to contribute to the nation's economic health. We raise three important issues here related to social justice.

First, when the rights discourse is braided together with the knowledge economy discourse, school issues are identified as the root cause as well as the fundamental solution of both educational inequality and economic decline. It is important to point out that there is a clear notion of social justice underlying this assumption. But it is a narrow and limited notion of justice as redistribution of school personnel who can boost achievement scores and thus boost the economic health of the nation. There is little or nothing in this braided discourse about the impact of contextual factors on achievement or the need for revised curriculum or instructional goals. There is also little or nothing about the justice of recognition, indicated by the lack of strategies for participation by all social groups in the discourse about what is fundamental in education, including those historically marginalized. As Joyce King (2006) has argued, "if justice is our objective" (p. 337) in education, then we must recognize the ways "ideologically distorted knowledge sustains societal *in*justice" (p. 337). Unfortunately, as King points out, when this is ignored, it can lead to the untenable situation in which "equal access to a faulty curriculum" (p. 337)—or, we would add, equal access to teachers governed by a faulty accountability system—is assumed to constitute justice. In addition, there is little or nothing in this braided discourse about the myriad complex factors, besides teacher quality, that heavily influence the economic health of the nation and its ability to compete in the knowledge economy, such as monetary, trade, and industrial policies, not to mention involvement in two major wars.

Second, one of the arguments underlying the braided discourse of the teacher quality gap and the knowledge economy is that in order to close the achievement gap, poor and minority students—who enter school "behind" (Peske & Haycock, 2006)—need effective teachers who help them gain access to the knowledge traditions and values of dominant groups. Here, the basic assumption is that diversity is a deficit and a barrier to be overcome, and the goal is assimilation into dominant culture and school-based ways of knowing. There is little recognition here of the need for teachers and schools to respect and utilize the resources students bring to school in order to build new knowledge and skills and broaden the curriculum. For example, there is no discussion about high quality teachers who incorporate the "funds of knowledge" of local communities and cultural groups (e.g., Moll, 2009), build on the language skills and cultural identities of students whose first language is not English (e.g., Brisk, 2007), and assume a "capacity framework" in working with students with special needs (El-Haj & Rubin, 2009).

Finally, it is clear that the ideology underlying this braided discourse focuses on the economic need to have a competitive work force, not on education as a fundamental human right. Perkins (2004) uses the term "corporatocracy" to refer to the alignment of business, government, and financial interests, and institutions in contrast to democracy. Building on Perkins, Sleeter (2008) argues that corporatocracy is aimed at consolidating global economic power for the benefit of the elite and is thus antithetical to the fundamental principles of democracy. What is particularly disturbing here is that there is virtually no discussion about the democratic goals of education—for example, the critical capacity of the American electorate, the level and quality of civic engagement in local and national arenas, or the opportunities of the populace to engage in meaningful work.

Example 2: National Accreditation Policies Regarding Teacher Education

The development of standards for the profession has been a central part of the movement to professionalize teaching and teacher education for three decades. Since the mid-1980s, the National Council for Accreditation of Teacher Education (NCATE) has evaluated teacher preparation programs seeking national accreditation in terms of the professional knowledge base and later the conceptual frameworks that shaped and connected coursework and fieldwork. In the early 2000s, NCATE initiated performance standards consistent with those of the National Board for Professional Teaching Standards and the Interstate New Teacher Assessment and Support Consortium (Darling-Hammond, Wise, & Klein, 1999). Consistent with the larger evidence-based education movement and with the shift from inputs to outcomes in terms of educational accountability, the NCATE standards required that programs provide "compelling evidence" (Williams, Mitchell, & Leibbrand, 2003, p. xiii) of teachers' content knowledge and performance and that programs have built-in data-driven assessment systems. Our second example of teacher education policy and social justice focuses on national accreditation policy, particularly the controversy that erupted about NCATE's alleged social justice requirement in 2005–2006 and subsequent developments.

NCATE's Alleged Social Justice Standard

Standard 1 of NCATE's (2002) standards stated that all teacher candidates must "know and demonstrate the content, pedagogical, and professional knowledge, skills, and dispositions necessary to help all students learn." NCATE's glossary of terms defined "dispositions" as follows:

> The values, commitments, and professional ethics that influence behaviors toward students, families, colleagues, and communities ... Dispositions are guided by beliefs and attitudes related to values such as *caring, fairness, honesty, responsibility, and social justice.* [emphasis added].
>
> (p. 53)

In order to meet this standard, some teacher education institutions developed assessments to document and measure candidates' dispositions.

Controversy about requiring certain dispositions for state teacher certification, particularly social justice, came to a head during the winter of 2005–2006, ignited by an article that appeared in the *Chronicle of Higher Education* and a request that the U.S. Department of Education investigate the propriety of NCATE's reference to social justice in its standards. However, this controversy, which has been the topic of a number of analyses, including our own (Cochran-Smith, 2006; Cochran-Smith, Barnatt, Lahann, Shakman, & Terrell, 2009; Cochran-Smith & Demers, 2007), has much deeper roots in the larger and ongoing culture wars in the U.S., and, more particularly, in the curriculum wars related to history, the social studies, and literature. These are connected to sharply contested ideas about the purposes of schooling in society, the politics of knowledge, and the current backlash against universities, which are characterized by some conservatives as hotbeds of radical thought and revolution. We briefly recap the denouement of this controversy below.

An article in the *Chronicle of Higher Education* headlined "We Don't Need That Kind of Attitude" and subtitled "Education schools want to make sure prospective teachers have the right disposition" (Wilson, 2005) described grievances pending at several institutions where teacher candidates were expected to acknowledge ideas such as "white privilege," agree to be "agents of change," and/or meet specific program criteria regarding their dispositions for social justice. The *Chronicle* story was picked up by many journalists, cartoonists, and bloggers, whose commentaries (e.g., Will, 2006) added to growing doubt about the value of the collegiate teacher education curriculum.

Just a month before the *Chronicle* article appeared, the National Association of Scholars (NAS) requested that the U.S. Department of Education investigate the "educational and constitutional propriety" of the reference in NCATE's standards to social justice, a term NAS argued was "necessarily fraught with contested ideological significance" (National Association of Scholars, November 2, 2005). The position of NAS was that teacher education programs should be based on "objective" standards and "core knowledge" rather than ideology. In the same letter, NAS also challenged standards at a school of social work, where the stated purpose was to prepare social workers to help alleviate poverty, oppression, and social injustice.

Some months later, and following intense political pressure and media attention, NCATE withdrew the language of social justice from its standards. In testimony before the Education Department's National Advisory Committee on Institutional Quality and Integrity, then NCATE President Arthur Wise stated:

> I categorically deny the assertion that NCATE has a mandatory 'social justice' standard. We don't endorse political and social ideologies. We endorse academic freedom, and we base our standards on knowledge, skills and professional disposition.
>
> (Powers, 2006)

It is worth noting that, although much of the NCATE debate seemed on the surface to be about the issue of professional dispositions, no public debate ensued about teacher candidates' dispositions toward caring, fairness, or honesty. There is little doubt that it was social justice that incited the critics.

Current Policy in National Accreditation Agencies

Since the brouhaha described above, NCATE has redefined dispositions and issued a "call to action" (National Council for Accreditation of Teacher Education, 2007) regarding social justice. Subsequently the organization has also revised its standards, which, according to current NCATE President James Cibulka, are much more rigorous (Lee, 2009). The social justice call to action begins with this statement:

> We, the members of the education profession, believe that high quality education is a fundamental right of all children … At least since *Brown* v *Board of Education* in 1954, our Nation has struggled to provide equal educational opportunity to all children. Now federal law requires that no child be left behind. Social justice demands that we take appropriate action to fulfill these promises by assuring high quality education for all children.

The statement emphasizes that NCATE's major commitment is to ensure that school students of all "races, ethnicities, disabilities/exceptionalities, and socioeconomic groups" have access to teachers who demonstrate "fairness in educational settings by meeting the educational needs of all students in a caring, non-discriminatory, and equitable manner." These statements are very consistent with the language and logic of NCATE's Standard 4, "Diversity," which elaborates the knowledge, skills, and dispositions teachers need to work effectively with diverse students, families, and schools. The NCATE statement on social justice concludes with this declaration: "When the education profession, the public and policymakers demand that all children be taught by well-prepared teachers, then no child will be left behind and social justice will be advanced."

NCATE also redefined professional dispositions, stressing that NCATE never had a standard related to social justice, a term that was "wildly" misunderstood by critics. Dispositions were defined as:

> Professional attitudes, values, and beliefs demonstrated through both verbal and non-verbal behaviors as educators interact with students, families, colleagues, and communities … The two professional dispositions that NCATE expects institutions to assess are fairness and the belief that all students can learn.

It was also pointed out that those who criticized NCATE for "caving" to its critics also misunderstood the "non-existent" social justice standard.

Teacher education currently has a second national accreditor, the Teacher Education Accreditation Council (TEAC). Federally approved as an accreditor only since 2003, this organization has a much shorter and less colorful history with regard to social justice than does NCATE. TEAC accredits teacher preparation institutions on the basis of valid and reliable evidence regarding the claims faculty make about their graduates. TEAC's first "quality principle" focuses on teacher candidates' learning to be competent, caring, and qualified teachers by acquiring subject matter and pedagogical knowledge as well as caring and effective teaching skills. One of three "cross-cutting themes" regarding candidate learning is "multicultural perspectives and accuracy." This is elaborated as: "Candidates must demonstrate that they have learned accurate and sound information on matters of race, gender, individual differences, and ethnic and cultural perspectives." TEAC's accreditation principles are much briefer and less specific than NCATE's standards. There is no mention of social justice or equity in TEAC's statement of accreditation principles, and the only reference to diversity has to do with the recruitment and retention of diverse teaching candidates and with programs' responsiveness to the demand for teachers in high need areas.

National Accreditation Policy and Social Justice

Currently NCATE and TEAC are involved in joint efforts to form one teacher education accrediting system for the nation that is unified in goals and voice, but that also offers two alternative pathways to accreditation—essentially one pathway will be NCATE's system, and the other will be TEAC's system (Murray & Cibulka, 2009; Murray & Wise, 2009). Presumably NCATE's standards and TEAC's quality principles will remain more or less intact, as described above. The presence and absence of social justice in the language and logic of teacher education policy at the level of national accreditation suggests many critical issues. We take up two of these here.

First, we want to unpack the critiques of social justice in NCATE's standards. Of central importance is what we have called "the ideology critique" (Cochran-Smith et al., 2009), which focuses specifically on the criteria and standards according to which prospective teachers are admitted into or barred from entering the profession. The central assumption animating the ideology critique is that professional accreditation can be and ought to be apolitical, value-free, and neutral when it comes to moral and ethical issues. This presumes, of course, that there is a choice in education—as in all social institutions—between politics and no politics and that it is possible to engage in program accreditation without being political. Our position here is that a neutral and value-free kind of accreditation is neither desirable nor possible. Education, teacher education, and professional accreditation are social institutions that pose moral, ethical, social, philosophical, and ideological questions. It is wrong-headed—and dangerous—to treat these questions as if they were value neutral and ideology free. Of course teacher education for social justice is political—it has to do with who has power and access to learning and life opportunities. All professional education—whether in law, nursing, or education—is value-laden and ideological rather than neutral and apolitical. Once the ideological basis of professional education is acknowledged, then it stands to reason that debates about accreditation policy need to address openly the difficult choices and trade-offs that all choices about values and ideology entail.

The second point we take up is closely related to the first. Currently, national teacher education accreditation policy is being shaped by the desire to avoid anything that appears to be ideological or values-related. TEAC simply omits any mention of social justice issues, although it encourages individual institutions to establish their own goals. Minimally, TEAC stipulates that all candidates should have "multicultural perspectives," but they frame this in terms of accuracy and as part of the liberal arts tradition, thus sidestepping the issue of social justice. For example, in response to an early draft of the TEAC inquiry brief of the University of New Hampshire (UNH), the home institution of the second author of this chapter, TEAC leaders pointed out that the UNH group had erroneously stated that "multi-culturalism" was one of TEAC's three cross-cutting themes: "TEAC uses the term 'multi-cultural perspectives' ... 'Multiculturalism' may have a somewhat different connotation" (Mosberg, 2006, p. 15). We presume the connotation TEAC eschewed was the idea of multiculturalism as a belief system, doctrine, or philosophical theory accepted by a certain group of people as authoritative or appropriate.

On the other hand, NCATE's latest statement about social justice intentionally echoes the language of No Child Left Behind (NCLB) legislation, equating social justice with providing a high quality education for all children and making no mention of social activism, teachers as change agents, or the need for educators to work with others to identify and challenge the structures and systems that perpet-uate inequity. This perspective is very similar to parts of the "teacher quality gap and educational inequality" discourse we discussed in Example 1 above. As we showed, the teacher quality gap discourse is based on a narrow view of justice as the redistribution of school-related resources, especially high quality teachers who can raise test scores. This discourse does not acknowledge the impact of other factors, such as poverty, institutional racism, and white normativity on school outcomes. Nor does it attend to the need to revise curricular and instructional goals based on recognition of the knowledge perspectives of marginalized groups. In the end, NCATE's current perspective on social justice is somewhat confusing. Their accreditation standards, especially Standard 4, emphasize teachers' compe-tence in drawing on the cultural and linguistic resources of all students and affirming the value of diversity. Yet NCATE's document explicitly spelling out its current stance on social justice reframes its commitment in the language of NCLB.

Example 3: Local Program Policy Regarding Teacher Education

Across the U.S., institutions that prepare teachers are now driven by efforts to document and measure the impact of their programs and the effect of program graduates on students' learning. Whereas there is general consensus among teacher educators that accountability is important, many concerns have been raised about narrow definitions of success that focus exclusively on how the K-12 students of program graduates perform on standardized tests (Cochran-Smith & Fries, 2005; National Council for Accreditation of Teacher Education, 2007; Sirotnik, 2004). There have also been criticisms of the lack of attention to other outcomes such as preparing teachers for diverse populations, teaching students to participate in a democratic society, ensuring equitable learning opportunities for all students, and working to make schools more caring and just (Cochran-Smith & Zeichner, 2005; Michelli, 2005; Oakes, 2004; Sleeter, 2005; Villegas & Lucas, 2001). Along these lines, at some institutions—particularly those that prepare teachers for urban schools and/or to work with marginalized school populations—educators are trying to conceptualize and measure equity and social justice as outcomes of teacher preparation. Our third example of teacher education policy outlines the work of one teacher education program that has social justice as its overarching goal in order to raise questions about teacher education policy and social justice at the local level.

Constructing Social Justice as a Teacher Education Outcome at Boston College

Boston College, the home institution of the first author of this chapter, is a medium-sized Jesuit university with an historical commitment to social justice. In teacher education, this commitment is based on recognition of significant disparities in the educational opportunities, resources, achievement, and outcomes between minority and majority groups and of the historical and systemic structures that perpetuate disparities. This is coupled with the position that teachers should be committed to the democratic ideal and to diminishing inequities in school and society by challenging inequitable educational structures and systems and also recognizing and respecting all social groups and their knowledge traditions.

Given that social justice is a complex and multifaceted matter, it follows that constructing social justice as an outcome of teacher education requires complex and multiple measures. As part of its work as a national site of the Carnegie Corporation's Teachers for a New Era (TNE) initiative, Boston College's interdisciplinary "Evidence Team" (ET) developed multiple assessments investigating the processes and outcomes of teacher education. Each of these has a major strand designed to get at this outcome, which we briefly describe below.

The ET developed a series of surveys (Entry, Exit, One-, Two-, and Three-years out) administered to all teacher candidates and graduates (Ludlow et al., 2008), which assessed perceptions, expectations, expected career trajectories, sense of preparedness for teaching, and, once in the classroom, reported practices and strategies (Enterline, Cochran-Smith, Ludlow, & Mitescu, 2008). Embedded in each survey is a 12-item Likert-response scale pertaining to teaching for social justice

(Ludlow et al., 2008). Scale items encompass a number of key ideas, including: high expectations and rich learning opportunities for all students; an asset-based perspective on the resources students and families bring to school; the importance of critical thinking in a democratic society; the role of teachers as advocates and agents for change; challenges to the notion of a meritocratic society; teaching as an activity related to teachers' deep underlying assumptions about race, class, gender, disability, and culture; and the idea that culture, equity, and race ought to be topics that are visible and spoken about in all aspects of the curriculum (Enterline et al., 2008).

The ET also developed the Qualitative Case Studies (QCS) project featuring 22 in-depth longitudinal case studies following candidates from entry into a one-year master's program and extending through four years of teaching (or a shorter time period if a participant left the program or teaching at an earlier point). This project focused on interrelationships among: candidates' entry characteristics, program experiences, evolving practices, the learning opportunities they created for students, and larger school and social outcomes. The QCS project was based on 15 in-depth interview protocols, each with a unique focus but also including specific questions about social justice. The project also featured an extensive, multi-part protocol for observing candidates/graduates at work in classrooms with a major section on social justice.

Embedded within certain QCS interviews was the "Teacher Assessment and Pupil Learning" (TAPL) protocol for evaluating teachers' assignments/assessments and students' work (Cochran-Smith, Gleeson, & Mitchell, 2010). The TAPL uses two data sources: an assignment or assessment used by a teacher as the culminating task for a unit of study, and a class set of students' work generated in response to that assignment/assessment. During an "internal evaluation" of these data, candidates assess their own work and the work of their students within an interview setting. During a separate "external evaluation," trained raters use the framework of "authentic intellectual work" (Newman, 1996) to score the intellectual quality and cognitive complexity of the assessments and the students' learning, because this framework was consistent with the idea of learning to teach for social justice (Gleeson, Cochran-Smith, Mitchell, & Baroz, 2008).

> A centerpiece of teacher education at Boston College is practitioner inquiry, or, classroom research conducted by teacher candidates about problems and issues that emerge from the intersections of theory and classroom practice in K-12 classrooms. All candidates are required to participate in inquiry seminars that coincide with full-time student teaching and to complete an inquiry paper. Each candidate poses a question, collects multiple sources of evidence and classroom data, analyzes and interprets the data in terms of immediate and ongoing decisions, and connects all of this to larger issues related to school and society. These inquiry projects are assessed according to candidates' research skills, the quality of their curriculum and instruction, their efforts to teach for social justice, and their students' learning.
>
> (Cochran-Smith, Barnatt, Friedman, & Pine, 2009)

Finally, the Pre-service Performance Assessment (PPA), which is required by the State of Massachusetts, was designed as a standards-based performance measure of candidates' teaching competencies during the full practicum experience. The PPA assesses five standards—planning curriculum and instruction, delivering effective instruction, managing classroom climate, promoting equity, and meeting professional responsibilities (Massachusetts Department of Education, 2005). The Boston College ET enhanced the state-mandated PPA (then deemed the PPA+) by elaborating on the "equity" standard to include teaching for social justice (i.e., "promotes equity and social justice") and two new additional standards (i.e., "assesses and promotes student learning" and "demonstrates an inquiry stance in daily practice"). Clinical faculty supervisors rate teacher candidates on the extent to which they provide evidence that they meet indicators for each of the seven standards.

Local Assessment Policy and Social Justice

As we noted in the introduction to this example, many teacher education programs around the country are involved in efforts to establish and maintain local policies and practices that assess their graduates and program's impact in order to meet the requirements of their accreditors and also to ascertain whether they live up to their own high standards. The intention across most of these efforts is to develop the capacity within teacher education programs to assess progress and effectiveness, shift accountability from external policy to internal practice, and generate knowledge that can be used both in local programs and, in some cases, more broadly. The Boston College teacher education program, formerly accredited by NCATE, shifted to TEAC accreditation in 2008, partly to build on the work it had begun as part of the TNE project. Its TEAC inquiry brief drew on the quantitative, qualitative, and mixed methods data described in the assessments above to support seven major claims about its teacher candidates, including the first: "[Teacher candidates] believe in and are committed to teaching for social justice, defined as improving the learning of all students and enhancing their life chances." The program was fully approved and accredited in 2009, and its social justice orientation was not problematic. However, there are a number of issues related to social justice and local teacher education policy. We take up three of these below.

The most important issue has to do with the paradox of establishing local assessment policies that focus on social justice. On one hand, constructing social justice as an outcome is to a certain extent buying into the prevailing viewpoint that teacher education is an intervention to be evaluated according to its results and outcomes. On the other hand, the intention at Boston College—and at other institutions making similar efforts—was not simply to buy into, but rather to interrupt, the assumptions driving the outcomes thrust of teacher education, including: the assumption that teacher preparation is an intervention that can be tracked directly to student outcomes, the assumption that students' academic achievement is the sole important outcome of schooling, and the assumption that test scores are the most important measure of students' learning. The intention with much of local social justice policy is to work both within and against the system—to be transformative and to challenge the inequities perpetuated by the status quo, but also to prepare teachers in a program that is closely enough aligned with the professional accountability system to be approved. There is a fine line here.

Second, as we have argued, teaching for social justice is a complex concept with many aspects and thus it requires a set of complex measures to assess it. No single assessment tells the whole story. The Boston College social justice beliefs scale, for example, focuses exclusively on beliefs and perceptions, but does not address classroom practice or content and pedagogical knowledge. Likewise the TAPL highlights the intellectual quality of the learning opportunities teachers provide, but does not account for teachers' beliefs, relationships with parents and colleagues, or advocacy for students. Each assessment is a piece within a larger picture and must be considered in terms of the trade-offs it involves. When multiple assessments are taken together, there is a much richer picture of learning to teach for social justice than any single assessment can generate.

Finally, there are many issues involved in constructing social justice as an outcome of teacher education that fuses ideology and accountability. It is sometimes allowed that this may be appropriate at a private (and Jesuit) university like Boston College, but not appropriate at public institutions. The assumption here is that social justice is inappropriate local policy because it has to do with values, beliefs, and ideals, which are assumed not to be the proper purview of public teacher education. However, our position here is that teaching is a profession with certain inalienable purposes, among them challenging the inequities in access and opportunity that curtail the opportunities of some to obtain a high quality education and, at the same time, recognizing the values and knowledge of marginalized social groups. From this perspective, teaching is a profession that—by definition—has social responsibilities that include challenging the barriers that constrain access to educational resources and, at the same time, challenging the cultural hegemony of curriculum, educational policy, and the arrangements and norms of schools. This means that learning to teach for social justice is integral to the very idea of learning to teach, and thus teaching for social justice is not an outcome only for those prepared at private universities, but a fundamental outcome of teacher preparation in general.

Teacher Education Policy and Social Justice: Directions Forward

It is difficult to make generalizations about teacher education policy and social justice in the U.S. As we argued at the beginning of this chapter, teacher education policy is developed and implemented by multiple agents and at multiple levels. The preparation of teachers continues to be the responsibility of the states, yet there has been sweeping federal legislation that has ushered in an era of unprecedented federal control, which is now expanding even further due to the unparalleled funds available in Race to the Top and other initiatives. States continue to approve university-based teacher preparation programs, yet some states have yielded much of their authority to national accrediting agencies because of lack of resources and personnel. In other states, stringent requirements control the content as well as the structure of university teacher preparation programs at the same time that there are no such requirements for alternate entry pathways. Meanwhile every local institution that prepares teachers has its own policies regarding the recruitment, preparation, assessment, and evaluation of teachers.

What does this add up to? As our first two examples indicate, at the level of state and federal policy regarding teacher quality and teacher education, there does indeed appear to be a justice agenda. This is clear in many speeches by both President Obama and Secretary of Education Duncan, where the emphasis is on guaranteeing that all students, including those in under-served schools, have access to high quality teachers, and the rhetoric of "no excuses" for the failure of marginalized students continues to be prevalent. In addition, the Race to the Top initiative is in large measure a race to establish state-wide data-based accountability systems that track all students and teachers. Unfortunately, as our first example emphasizes, underlying this agenda is the assumption that school factors—especially teachers, principals, and union contracts—are responsible for achievement disparities and that acknowledging contextual factors constitutes an "excuse" for poor teacher performance along with tacit support for a failed system. The continued prominence of the "no excuses" theme suggests that teacher quality policies in the U.S. are operating from a narrow view of justice, which focuses only on the distribution to low-performing schools of teachers who can up test scores. Unfortunately, this ignores both the unjust distribution of broader access, power, and opportunity and—just as bad—also ignores the unjust omission from curricula and educational goals of the knowledge traditions and assets of diverse social groups.

Meanwhile, in some local programs scattered across the country, such as the program at Boston College, there are efforts to prepare teachers to work for social justice in urban and other under-served schools and to establish local policies that assess program success in terms of these efforts. Despite these pockets of change, the reach of local programs is limited, definitions of social justice vary from institution to institution, and all programs must work within a now firmly entrenched accountability system that treats student test scores as the bottom line. The discourse about social justice in local teacher education policy has been marginalized to a great extent.

References

American Recovery and Reinvestment Act. (2009). Retrieved from http://www.recovery.gov

Bacchi, C. (2000). Policy as discourse: What does it mean? Where does it get us? *Discourse: Studies in the Cultural Politics of Education 21*(1), 45–57.

Ball, S. (2008). *The education debate: Policy and politics in the twenty-first century*. Bristol, England: The Policy Press.

Brisk, M. (Ed.). (2007). *Language, culture, and community in teacher education*. Mahwah, NJ: Lawrence Erlbaum Publishers.

Cheek, J. & Gibson, T. (1997). Policy matters: Critical policy analysis and nursing. *Journal of Advanced Nursing, 25*, 668–672.

Cochran-Smith, M. (2006). Teacher education and the need for public intellectuals. *New Educator 2*, 1–26.

Cochran-Smith, M. (2010). Toward a theory of teacher education for social justice. In M. Fullen, A. Hargreaves, D. Hopkins, & A. Liberman (Eds.), *The international handbook of educational change*, 2nd edition. New York, NY: Springer.

Cochran-Smith, M., Barnatt, J., Friedman, A., & Pine, J. (2009). Inquiry on inquiry: Practitioner research and students' learning. *Action in Teacher Education*.

Cochran-Smith, M., Barnatt, J., Lahann, R., Shakman, K., & Terrell, D. (2009). Teacher education for social justice: Critiquing the critiques. In W. Ayers, T. Quinn, & D. Stovall (Eds.), *Handbook of Social Justice in Education* (pp. 625–639). Philadelphia, PA: Taylor and Francis.

Cochran-Smith, M. & Demers, K. (2007). Teacher education as a bridge? Unpacking curriculum controversies. In M. Connelly, M. He, & J. Phillion (Eds.), *The SAGE handbook of curriculum and instruction* (pp. 261–281). Los Angeles, CA: SAGE.

Cochran-Smith, M. & Fries, K. (2001). Sticks, stones, and ideology: The discourse of teacher education. *Educational Researcher 30*(8), 3–15.

Cochran-Smith, M. & Fries, K. (2005). Researching teacher education in changing times: Politics and paradigms. In M. Cochran-Smith & K. Zeichner (Eds.), *Studying teacher education: The report of the AERA panel on research and teaching* (pp. 69–110). Mahwah, NJ: Lawrence Erlbaum Associates Inc.

Cochran-Smith, M. & Fries, K. (2011). Teacher education and teacher quality: Policy-as-discourse. In A. Ball & C. Tyson (Eds.), *Studying diversity in teacher education*. Washington, DC: American Educational Research Association.

Cochran-Smith, M., Gleeson, A., & Mitchell, K. (2010). Teacher education for social justice: What's pupil learning got to do with it? *Berkeley Education Review X*(X), 35–61.

Cochran-Smith, M. & Zeichner, K. (Eds.). (2005). *Studying teacher education: The report of the AERA panel on research and teacher education*. Mahwah, NJ: Lawrence Erlbaum Associates Inc.

Darling-Hammond, L., Wise, A., & Klein, S. (1999). *A license to teach: Raising standards for teaching*. San Francisco, CA: Jossey-Bass.

Duncan, A. (2009a, May 5). Press Release: Education secretary launches national discussion on education reform. Retrieved from http://www.ed.gov

Duncan, A. (2009b). Press release: Remarks of Arne Duncan to the National Education Association—Partners in reform. Retrieved from http://www.ed.gov

Duncan, A. (2009c). States will lead the way toward reform: Secretary Arne Duncan's remarks at the 2009 Governor's Education Association. Retrieved from http://www.ed.gov

Duncan, A. (2009d). Teacher preparation: Reforming the uncertain profession. Remarks made at Teachers College, Columbia University, October 22. Retrieved from http://www.ed.gov

Earley, P. (2000). Finding the culprit: Federal policy and teacher education. *Educational Policy 14*(1), 25–39.

Education Equality Project. (2008). Home website. Retrieved from http://www.educationequalityproject.org

Education Equality Project. (2009). On improving teacher quality. Retrieved from http://www.educationqualityproject.org

El-Haj, T. & Rubin, B. (2009). Realizing the equity-minded aspirations of detracking and inclusion: Toward a capacity-oriented framework for teacher education. *Curriculum Inquiry, 39*(3), 435–462.

Enterline, S., Cochran-Smith, M., Ludlow, L., & Mitescu, E. (2008). Learning to teach for social justice: Measuring changes in the beliefs of teacher candidates. *New Educator 4*, 1–24.

Fischer, F. & Forester, J. (Eds.). (1993). *The argumentative turn in policy analysis and planning.* Durham, NC: Duke University Press.

Fraser, N. (2003). Social justice in an age of identity politics: Redistribution, recognition and participation. In N. Fraser & A. Honneth (Eds.), *Redistribution or recognition: A political-philosophical debate* (pp. 7–109). London: Verso.

Fraser, N. & Honneth, A. (Eds.). (2003). *Redistribution or recognition: A political-philosophical debate.* London: Verso.

Gewirtz, S. & Cribb, A. (2002). Plural conceptions of social justice: Implications for policy sociology. *Journal of Education Policy 17*(5), 499–509.

Gleeson, A., Cochran-Smith, M., Mitchell, K., & Baroz, R. (2008). Teacher education for social justice: What's pupil learning got to do with it? Paper presented at the Annual Meeting of the American Educational Research Association, New York.

Haycock, K. (2004). The real value of teachers: If good teachers matter, why don't we act like it? *Thinking K-16: A publication of the Education Trust 8*(1).

Hess, F., Rotherham, A., & Walsh, K. (2004). *A qualified teacher in every classroom: Appraising old answers and new ideas.* Cambridge, MA: Harvard Education Press.

Hogwood, B. & Gunn, L. (1990). *Policy analysis for the real world.* Toronto, Canada: Oxford University Press.

Honneth, A. (2003). Redistribution as recognition: A response to Nancy Fraser. In N. Fraser & A. Honneth (Eds.), *Redistribution or recognition: A political-philosophical debate* (pp. 110–197). London: Verso.

Howe, K. (1997). *Understanding equal educational opportunity: Social justice, democracy, and schooling.* New York: Teachers College Press.

Joshee, R. (2007). Opportunities for social justice work: The Ontario diversity policy web. *Journal of Educational Administration and Foundations 18*(1/2), 171–199.

Joshee, R. (2009). Multicultural education policy in Canada: Competing ideologies, interconnected discourses. In J. Banks (Ed.), *The Routledge international companion to multicultural education* (pp. 96–108). New York: Routledge.

Joshee, R. & Johnson, L. (2005). Multicultural education in the United States and Canada: The importance of national policies. In N. Bascia, A. Cumming, A. Datnow, K. Leithwood, & D. Livingstone (Eds.), *International Handbook of Educational Policy* (pp. 53–74). New York: Springer.

King, J. (2006). If our objective is justice: Diaspora literacy, heritage knowledge, and the praxis of critical studying for human freedom. In A. Ball (Ed.), *With more deliberate speed: Achieving equity and excellence in education—Realizing the full potential of Brown v. Board of Education* (pp. 337–357). Chicago, IL: University of Chicago Press.

King, J. (2008). Critical and qualitative research in teacher education: A blues epistemology for cultural well-being and a reason for knowing. In M. Cochran-Smith, S. Feiman-Nemser, J. McIntyre, & K. Demers (Eds.), *Handbook of research on teacher education: Enduring questions in changing contexts*, 3rd edition (pp. 1094–1136). New York: Routledge/Taylor & Francis.

Lee, S. (2009). Raising the bar on teacher education. Inside Higher Education. Retrieved from http://www.insidehighered.com

Ludlow, L., Pedulla, J., Enterline, S., Cochran-Smith, M., Loftus, F., Salomon-Fernandez, Y., and Mitescu, E. (2008). From students to teachers: Using surveys to build a culture of evidence and inquiry. *European Journal of Teacher Education 31*(4), 1–19.

Luke, A. (2002). Beyond science and ideology critique: Developments in critical discourse analysis. *Annual Review of Applied Linguistics 22*, 96–110.

Massachusetts Department of Education. (2005). Preparing educators: Pre-service performance assessment. Retrieved from http://www.doe.mass.edu

McKinsey & Company. (2009). *The economic impact of the achievement gap in America's schools.* Washington, DC: Author.

Michelli, N. (2005). The politics of teacher education: Lessons from New York City. *Journal of Teacher Education 56*(3), 235–241.

Moll, L. (2009). *6th Annual Brown Lecture: Mobilizing culture, language and education practices: Fulfilling the promises of Mendez and Brown.* Washington, DC: American Educational Research Association.

Mosberg, L. (2006). Reviewer's comment: Draft of UNH Inquiry Brief.

Murray, F. & Cibulka, J. (2009). More rigorous requirements for teacher education will encourage programs to emphasize clinical training, focus on critical needs of P-12 schools. Retrieved from http://www.ncate.org

Murray, F. & Wise, A. (2009). Towards a unified accreditation system for educator preparation. Retrieved from http://www.teac.org

National Academy of Education. (2009). *White Paper: Teacher quality and distribution.* Washington, DC: Author.

National Association of Scholars. (2005). Letter to Assistant Secretary for Post Secondary Education, United States Department of Education, November 2.

National Center on Education and the Economy. (2006). *Tough choices or tough times: The report of the new commission on the skills of the American workforce (executive summary).* San Francisco, CA: Jossey-Bass.

National Commission on Excellence in Education. (1983). *A nation at risk: The imperative for educational reform.* Washington, DC: U.S. Government Printing Office.

National Council for Accreditation of Teacher Education. (2007). Call to action by NCATE's executive board on October 27. Retrieved from http://www.ncate.org

National Council for Accreditation of Teacher Education. (2002). *Professional standards for the accreditation of schools, colleges, and departments of education.* Washington, DC: Author.

National Governors Association, Council of Chief State School Officers, & Achieve, Inc. (2008). *Benchmarking for success: Ensuring U.S. students receive a world-class education.* Washington, DC: National Governors Association.

Newman, F. A. (1996). *Authentic achievement: Restructuring schools for intellectual quality.* San Francisco, CA: Jossey-Bass.

North, C. (2006). More than words? Delving into the substantive meaning(s) of "social justice" in education. *Review of Educational Research 76*(4), 507–536.

Oakes, J. (2004). Investigating the claims in *Williams v. State of California*: An unconstitutional denial of education's basic tools? *Teachers College Record 106*(10).

Oakes, J., Lipton, M., & Renee, M. (2006, July). Research as a tool for democratizing education policymaking. Paper presented at the International Invitational Symposium on Figuring and Re-configuring Research, Policy and Practice for the Knowledge Society, Dublin, Ireland.

Perkins, J. (2004). *Confessions of an economic hit man*. San Francisco, CA: Berrett Koehler Publishers.

Peske, H. & Haycock, K. (2006). *Teaching inequality: How poor and minority students are short-changed on teacher quality: A report and recommendations by the Education Trust*. ED 494820-ERIC Document.

Popkewitz, T. & Lindblad, S. (2000). Educational governance and social incluson and exclusion: Some conceptual difficulties and problematics in policy research. *Discourse: Studies in the Cultural Politics of Education 21*(1), 5–44.

Powers, E. (2006). A spirited disposition debate. *Inside Higher Education*. Retrieved from http://www.insidehighered.com

Reich, R. (2002). *Bridging liberalism and multiculturalism in American education*. Chicago, IL: Chicago University Press.

Sharp, L. & Richardson, T. (2001). Reflections on Foucauldian discourse analysis in planning and environmental research. *Journal of Environmental Policy and Planning 3*(3), 193–210.

Sirotnik, K. (2004). *Holding Accountability Accountable*. New York: Teachers College Press.

Sleeter, C. (2005). *Unstandardizing Curriculum: Multicultural Teaching in the Standards-Based Classroom*. New York: Teachers College Press.

Sleeter, C. (2008). Teacher education, neoliberalism, and social justice. In W. Ayers, T. Quinn, & D. Stovell (Eds.), *The handbook of social justice in education* (pp. 611–624). Philadelphia, PA: Taylor & Francis.

Stone, D. (1997). *Policy paradox: The art of political decision making*. New York: Norton.

Stone, D. (2002). *Policy paradox: The art of political decision making*, revised edition. New York: Norton.

Vagelos, R. (2006). Rising above the gathering storm: Energizing and employing America for a brighter economic future. *U.S. Senate's Subcommittee on Education and Early Childhood Development*. Retrieved from http://www.nationalacademies.org

Villegas, A. M. & Lucas, T. (2001). *Preparing culturally responsive teachers: A coherent approach*. Albany, NY: SUNY Press.

Will, G. (2006). Ed schools vs. education. *Newsweek*, January 16. p. 98.

Williams, B., Mitchell, A., & Leibbrand, J. (2003). *Navigating change: Preparing for a performance-based accreditation review*. Washington, DC: National Council for Accreditation of Teacher Education.

Wilson, R. (2005). We don't need that kind of attitude. *Chronicle of Higher Education 52*(17), A8.

Wineburg, M. (2006). Evidence in teacher preparation: Establishing a framework for accountability. *Journal of Teacher Education 57*(1), 51–64.

Young, I. (1990). *Justice and the politics of difference*. Princeton, NJ: Princeton University Press.

EDITORS' COMMENTARY

The chapter by Cochran-Smith and Fries (Chapter 9) reminds us that there is a fundamental and important purpose of schooling in the United States. It is to assure equal opportunities for all children and to challenge "practices that reinforce inequities." We note that there is no explicit or implicit reference to social justice in five of the six policy case studies and two reflective essays presented in the earlier chapters.[1] One might speculate that the authors believed it was more important to raise other issues: the nature of the federal role, the challenges posed to an independent standards board, the shifting sands of teacher education accountability, and the lack of an empirical base for policy decisions. Another explanation is that the weight of debates over the how and when of accountability is so great that it pushes other issues into the deep recesses of the policy agenda. Or, it may be that, as Cochran-Smith and Fries suggest, the teacher education establishment has aligned itself with "the powerful knowledge economy discourse, which dominates many federal and state policy debates about education and the now firmly-entrenched system of accountability that is driving it." As such economic necessity seems to trump civic engagement. All are reasonable theories, but the result is the same: issues of social justice and the important conversations that must occur about what it means to create and live in a just society are subdued to the point of near silence.

The case studies of Florida and Louisiana accountability policies reflect an emphasis on teacher quality and the knowledge economy. To achieve an education system that produces individuals with an orientation exclusively attuned to building economic prowess, decision makers seem to buy into the logic that academic measurements must be standardized. The theory of sense making co-construction as described by Datnow and Park suggests that implementation of policy is shaped by local context. A consequence of attempts to standardize evaluations of teachers and the programs that prepare them is restricted freedom for teachers and teacher educators to react to local contexts. That is, teacher education faculty members in a college or university may hold a common expectation that new teachers internalize social justice into their practice; however, state testing and other accountability measures may not be framed to be compatible with a social justice approach. Cochran-Smith and Fries discuss the challenge at Boston College to meld their assessments with those required by the State of Massachusetts. This is difficult work and, given a fractious policy environment in some states, may not be a challenge teacher educators can assume when struggling to justify the existence of their programs. When policy mandates appear to be ill conceived and overwhelming, education deans and faculty may turn to their accreditation agency or professional societies to speak out on their behalf. However, Cochran-Smith and Fries assert that NCATE and AACTE have aligned with those who are promoting an agenda linking teacher education to the promotion of a knowledge economy. Whereas this should not preclude a strong stance on social justice, the authors suggest this has been the case.

In Chapter 5, Kenneth Zeichner discusses the expense associated with sophisticated evaluations of teachers, and we assume this also is the case for observational instruments developed at Boston College. If this is so, can these instruments be generalized for use in other teacher preparation programs? And, related to that, what would be the expense to use them?

Discussion Questions

1. The authors of this chapter discuss social justice within the context of collegiate-based teacher education programs. Can, and do, truncated alternative

routes to teacher certification prepare educators instilled with dispositions to promote equity?

2. Will the current climate in the United States, which seems to be accepting of those who advocate intolerance, further challenge efforts to imbed social justice in the teacher preparation curriculum?

3. The authors cite Deborah Stone, who writes: "Each idea is an argument ... in favor of different ways of seeing the world ... There are multiple understandings of what appears to be a single concept, how these understandings are created, and how they are manipulated as part of political strategy." Given the clash of values and ideas reflected in the five discourses described by Cochran-Smith and Fries, is there a way to find common ground and understandings among them?

Note

1 Chapter 7 (by Cutler, Alvarez, and Taylor) describes an emphasis in their K–16 partnership.

10

EDITORS' REFLECTIONS

MAKING SENSE OF TEACHER EDUCATION POLICY AND MOVING TOWARD A COMMON WORLD VIEW

Nicholas M. Michelli

How does one make sense from these descriptions of policy, analyses of policy, reports of outcomes, hopes, and frustrations for enhancing teacher education? Are there ways of thinking about them that help us understand the polices better and perhaps influence them? As I talked with authors and read through these chapters, I felt as though I was reliving my 25 years as a dean both at Montclair State University in New Jersey and then as University Dean for Teacher Education for the City University of New York. I thought about my eight years on the New York State Professional Standards and Practices Board and my eight years as Chair of AACTE's Combined Committee for Governmental Relations, which included meeting with members of Congress on important educational issues and proposed legislation several times a year. I realized with glee that as a faculty member in the Ph.D. program in Urban Education at CUNY's Graduate Center I now have the luxury of writing about policy and its impact instead of losing sleep over it—but, unfortunately, I still lose sleep.

As I begin, I think it is important to note that policy is made in a political context, and policymakers all have world views that drive their decisions. Lakoff, in *Moral Politics* (Lakoff, 2002), uses metaphors to help us understand the different perspectives that make up these world views. Whether we call them liberal and conservative, or nurturing mother and strict father—he gives us a way of understanding the bases for these metaphors and the source of the conflict that is evident in the policy world. I use Lakoff to augment the idea of sense-making about policy. I believe one cannot make sense of policy without trying to reconstruct the world view from which it comes. Stone's notion of policy paradox helps to illustrate the potential for unintended consequences. Then I think about the very helpful analysis by the Northwest Regional Educational Labs, *Toward a Research Agenda: Understanding and Improving the Use of Research Evidence* (Nelson, Leffler, & Hanson, 2009). That study was specifically aimed at analyzing how policymakers make their judgments and how research fits into the process. Their findings were predictable:

> Among our findings, one of particular importance to researchers stood out: In our study policymakers and practitioners did not mention research evidence as often, nor discuss it as strongly, as other sources of information. Study participants expressed skepticism about research evidence (empirical findings derived from systematic methods and analyses) and noted its limitations. While almost all participants stated that research evidence plays a part in policy and practice decisions, they rarely identified it as a primary factor, most study participants responded that research evidence played a more indirect or secondary role.
>
> (Nelson et al., 2009, p. iii)

Their mistrust of research, they say, comes from,

> their own lack of sophistication in acquiring, interpreting, and applying research. They also cited obstacles such as time constraints, the volume of research evidence available, the format in which it is presented, and the difficulty in applying research evidence to their own situations.
>
> (Nelson et al., 2009, p. iv)

What did make a difference in their judgments?

> policymakers and practitioners regard evidence as a key factor in decision-making, but they take a more pragmatic approach to acquiring and using it. They define evidence broadly as local research, local data, personal experience, information from personal communications, gut instinct or intuition, and the experience of others, in addition to research evidence. In fact, focus group members and interviewees did not draw a distinction between research evidence and general evidence derived from these other sources.
>
> (Nelson et al., 2009, p. iv)

Who do they listen to in making their judgments?

> A central feature to the research utilization process was the role of intermediaries. Throughout our focus group discussions and interviews, participants repeatedly referred to their reliance on intermediaries, who were

described as unbiased organizations and individuals that can help locate, sort, and prioritize the available research. Intermediaries include research institutions, professional organizations, partners, coalitions, networks, peers, and constituents. Within these intermediary organizations, policymakers and practitioners appear to have a special relationship with small groups of "trusted individuals," who are valued as credible, objective sources of information.

(Nelson et al., 2009, p. v)

None of this should come as a surprise to any of us who have sought to influence policymakers, whether in local, state, or national contexts. This puts academics, who presumably rely on research evidence for judgments, at a disadvantage, and may give us some guidance for how to approach influencing policy while we find a way to communicate research better and work to be among those "trusted individuals" who are listened to. One paradox for me, given this aversion to complicated research, is how so technical a process as value-added assessment has come to have such credibility and to be embraced with such certainty. This issue of certainty (and I have seldom seen policymakers quite so certain) is explored in a paper by Charles Manski, of the Department of Economics and Institute for Policy Research at Northwestern University. The title "Policy Analysis with Incredible Certitude" gives some sense of Manski's content and conclusions (Manski, 2010). He concludes that policymakers like certainty, and that certainty requires strong assumptions—the stronger the assumption, the more certitude, and the more likelihood of a wrong outcome. In Zeichner's chapter (Chapter 5), his quotation from Mencken illustrates this—often we turn to solutions to complex problems that are quick, easy, and wrong. Manski uses the following example to illustrate the incentives policymakers have to pushing to certitude—no alternative assumptions:

> The scientific community rewards those who produce strong novel findings. The public, impatient for solutions to its pressing concerns, rewards those who offer simple analyses leading to unequivocal policy recommendations. These incentives make it tempting for researchers to maintain assumptions far stronger than they can persuasively defend, in order to draw strong conclusions.
>
> The pressure to produce an answer, without qualifications, seems particularly intense in the environs of Washington, D.C. A perhaps apocryphal, but quite believable, story circulates about an economist's attempt to describe his uncertainty about a forecast to President Lyndon B. Johnson. The economist presented his forecast as a likely range of values for the quantity under discussion. Johnson is said to have replied, "Ranges are for cattle. Give me a number."

(Manski, 2010, p. 3)

I conclude that the current fascination with the value-added metric is the result of the need for a clear and certain measure for making judgments, especially in times of declining budgets. Of course Zeichner points out just how expensive the system must be in order to have anything close to reliable data, and Fleener and Exner's chapter (Chapter 3) questions the utility of the value-added data in Louisiana.

Here is another way to look at it, and clearly it is a reflection of my world-view—which in this case focuses on what I believe the purposes of education should be in a democracy. From my perspective, democracy and social justice are intertwined and certainly I embrace the social justice implications so well presented in the Cochran-Smith and Fries chapter (Chapter 9), in which they argue that all education has a social justice element. They call for excellent teachers for all students, not just some, preparation for the knowledge economy, and addressing the teacher quality gap and educational inequality as matters of social justice. To that I would add the importance of teachers helping students learn to imagine their possibilities—embracing Maxine Greene's idea that "We can't become what we can't imagine" (Greene, 2001). How can these purposes of education be accounted for in a predominantly value-added system?

Our title for this book is *Teacher Education Policy in the United States: Issues and Tensions in an Era of Evolving Expectations.* This pushes me to think about the direction in which the expectations are evolving. So far as I can tell from reading these chapters and participating in the study of policy, the expectations are rising in a very narrow and, I might add, easily measured area of education—namely, student performance on standardized tests. I don't dismiss the role standardized tests can play, although I doubt along with others, including Hess (Chapter 2), if we can rely on them to make high-stakes judgments about educators' careers. In addition, I would argue for the rising expectations that are reflected in these questions:

- Are more students graduating from high school? What is the effect of standardized tests on school leaving, that is, students leaving school without graduating?
- Is the graduation rate about equal when sorted by whatever diversity factor we want to use?
- Are we succeeding in diversifying the teaching force?
- Are the achievement gaps narrowing?
- Are graduates leading rich and rewarding personal lives? Are they going to college? Are they moving to the highest place in the economy they are capable of? Where are they five years after leaving school?
- Does accreditation eliminate weak programs? Is it supposed to?
- We know that school climate can be measured as well as of its effect on teacher attrition and student learning (Cohen et al., 2009). What is the effect on school climate when we rely on value-added measures?

I am not arguing for less accountability; in fact, I am arguing for more. If we really want to affect schools, I would advocate spending our limited assessment resources on answering these questions, probably along with value-added measures that primarily aggregate the impact of large numbers of teachers on test scores. Teachers are indeed in part accountable for improving test scores as one measure of why we educate, but educators should also be held accountable for a role in addressing these other areas where we need improvement—recognizing that educators alone cannot solve them all. Addressing these questions really would represent an evolution to rising expectations.

I understand that some will say that better teachers who raise student scores will fix these problems. Actually, from my view, it is a policy paradox overall that pressure on principals and then on teachers to raise standardized test scores, or whatever "value-added" measure might be chosen, may well depress the gradua-tion rate by increasing school leaving. How might a 9th-Grade student in an urban school perceive the utility of education when he or she cannot imagine passing a standardized test or sustain being subjected to years of irrelevant test preparation to raise scores? And isn't it a paradox that lowering graduation rates will be dispro-portionately borne by black and Latino children just as we approach a time when we will no longer have any majority in this country? And, under these circum-stances, fewer students of color will go to college, so further diversifying the teaching force is less likely to be achieved. Given the pressures and their negative effects on school climate, more teachers may well leave. The achievement gap may narrow, but only for those who stay in school. A connection between these pres-sures and school leaving demonstrably leads to higher incarceration rates, less earning power, more welfare, and poorer health. In fact, one of my doctoral students has made a compelling argument that the school-leaving rate is a public health problem (Ruglis, 2009). Perhaps if school leaving were seen as a public health problem, policymakers might notice. Incarceration, earning power, welfare, and health are some very expensive problems for our society. These problems/ issues are not someone else's issues. We must recognize the connection between completing high school and these issues: that they are easily related to how we prepare teachers, why we think schools exist, and how much we really care. I am not suggesting that teacher education policy alone can solve all these problems, but that the connection between having excellent teachers and staying in school is demonstrable. To the extent that we ignore these questions in pursuit of simpler to measure outcomes—test scores—the direction of the evolution of expectations is downward.

Can a broader perspective on why we educate, a different world view, coexist with value-added measures? I am having some doubts based on my own experience. In the early part of the last decade my office put up $300,000 to begin planning and carrying out an evaluation of different pathways into teaching in New York City and put out an RFP (Request for Proposal). The group we selected to carry out the work and answer the questions—all of which were policy questions—included a distinguished group of labor economists and educators then at SUNY Albany and Stanford University. The result was the Pathways Study,[1] which with enormous support from foundations, including Carnegie and Spencer, is still examining the questions we raised using a value-added measure as one piece of the puzzle. The value-added measure in this case was not used to assess individual teachers, or to assess colleges as occurred in Louisiana. It was used to assess pathways into teaching across all the institutions serving New York City to find out which one achieved better results by these measures. A piece of this research for those of us working in the CUNY system was to gather data on not only our impact on standardized tests—important to policymakers and to us—but also to develop extensive qualitative data on our success in preparing teachers who focus on imagination, teach for democratic behaviors, and address school violence and bullying and other elements of an expanded view of the purposes of education in a democracy. I found the early data convincing. Graduates of traditional college-based pathways which did much more than focus on standardized tests did as well or better as other pathways on the value-added measures (that is a generalization, but essentially accurate—there were some nuances between mathematics and literacy scores). When I went to policymakers in New York, at every level, and discussed these data their response set me back. They said, "Well, if you weren't trying to do all those other things, imagine how high the standardized test scores could be." I also know policymakers who think that the kind of partnership described in the Cutler, Taylor, and Alvarez chapter (Chapter 7) diminishes the potential for growth in test scores. They fail to recognize the critical importance of seeing not only faculty in education, but also those in arts and science and in P-12 schools as collaborating teacher educators. I know many policymakers who have no idea that there might be a social justice goal for education. That experience I had with the Pathways study illustrated the differences in world view more than any other experience.

Let me turn briefly to Frederick Hess's piece in our chapter on federal influence (Chapter 2), and offer some advice. The Emihovich, Dana, Vernetson, and Colón chapter (Chapter 4) gives us a good sense of the politics of Florida's Senate Bill 6 and the round after round of changes in that state in pursuit of teacher assessment. I agree with Hess's analysis. Student performance should be one element of what we look at in assessing teachers, although I am not sure we would agree on what else to include besides standardized tests or on the appropriate relationship of the measures to teacher pay and other tools for change.

There is another part of what Hess writes that I want to emphasize. He suggests that we (I think we are the professionalizers he refers to) fail to recognize the "good intentions" of advocates for reforms, including those who advocate using teacher pay in the manner he suggests. He also poses the following question:

> Will the next decade be another decade of line-in-the-sand declarations and dueling bouts of moral posturing, or will it be about how we use the mundane tools of data, termination, evaluation, markets, professional norms, training, and the rest to finally make teaching start to look like a profession in the modern era?

Of course we start with agreement on what measures should go into deciding to use those "mundane" tools, or even whether they are the right tools. But Hess's point about eliminating moral posturing and doubting good intentions can be a useful starting place for a meaningful conversation. Starting that conversation requires that the "critics" (sometimes called the "deregulators") recognize that the "professionalizers" also have good intentions and that the differences we have are a matter of world view rather than of right and wrong. Engaging in real dialogue that recognizes the good will of all, trying to understand and emphasize our different world views and resulting positions, listening carefully to alternative arguments rather than dismissing them outright, and compromising in the interests of children up to the point where we maintain our integrity may be the most important hallmarks of democratic living.

Is there room for merging the world views? Can research of the sort Zeichner suggests, or of the sort Cochran-Smith has carried out at Boston College as part of the Teachers for a New Era project, be considered? Can the implications of The Pathways Study that are often overlooked by policymakers get into the mix? Can instruments like PACT give us useful in-class data on teachers? And I might add along these lines that others, including the State of Wisconsin under the leadership of Francine Tompkins of the University of Wisconsin System in partnership with the American Association of State Colleges and Universities, is developing performance instruments that can yield very useful data. In this work, 16 public and private colleges and universities, including faculty in education, arts and science and P-12 schools, have worked for three years to agree on the observable classroom performance in clinical experiences that can best predict the future success of preservice mathematics and science teachers. Ultimately there will be a common set of indicators used across all colleges in the state and used for program improvement. The state has guaranteed that comparisons across institutions will not be used in any pejorative way and will not be released publicly.[2]

If we can agree on some merger of world views and on how best to use research, maybe we can move forward. Like Zeichner in his chapter, I argue that we must rise above the bickering and "above our own self-interest and learn to work in more productive ways with those who hold positions different from our own."

If we are to do this, then we have to take a new look at and seek a new relationship with the critics of teacher education. I am reminded of a story I heard years ago about Adlai Stevenson's run for the presidency in 1952. A supporter is reported to have said to him, "Mr. Stevenson, you will surely win because every intelligent, thinking, caring American will vote for you." Stevenson replied, "That's not enough. I need a majority." It is an amusing story (and we know he didn't need a majority of the electorate), but think about the implications. Those who agree with Stevenson (or perhaps teacher educators) are seen as the intelligent, thinking, caring ones. What about those who don't agree with teacher educators? Are they all not intelligent, unthinking, and uncaring? And, what about those who disagree with the critics? We all have to change the way we perceive differences and each other if we stand a chance of moving toward the kind of educational system we want for our children, but I think it can only happen if we have a "big tent" approach where multiple prospectives are included, respected, and examined, put aside the anger and emotions, and see where it leads.

Notes

1 For background on The Pathways Study and the research reports to date, go to www. teachpolicyresearch.org. The primary researchers on this work are James Wyckoff, William Boyd, Hamilton Lankford, Pamela Grossman, and Susanna Loeb.
2 For more information on the Wisconsin Project, see http://tqi.uwsa.edu/fipse/index.htm.

References

Cohen, J., Michelli, N., & Pickerel, T. (2009). School climate: Research, policy, practice and teacher education. *Teachers College Record 111* (1, January), 180–213.
Greene, M. (2001). *Blue guitar.* New York: Teachers College Press. Also in personal correspondence with the author.
Lakoff, G. (2002). *Moral politics.* Chicago: University of Chicago Press.
Manski, R. (2010). *Policy analysis with incredible certitude.* Northwestern University, Evanston, IL.
Nelson, S., Leffler, J., & Hanson, B. (2009). *Toward a research agenda: Understanding and Improving the use of research evidence.* Portland OR: Northwest Regional Educational Laboratory.
Ruglis, J. (2009). Death of a dropout: (Re)theorizing school dropout and schooling as a social Determinant of Health. Unpublished Ph.D. dissertation. New York: The Graduate School, City University of New York.

THE FUTURE OF TEACHER EDUCATION

David G. Imig

For the past 30 years I have been engaged in a form of sense-making. I have spent a career trying to make sense of policy pronouncements and political actions, economic analyses and social developments, demographic shifts and normative changes, and the way they affect collegiate-based teacher education. I spent that time (a lot of it before the invention of the internet) sorting through and reflecting on the actions of a host of policymakers and policy informers. I spent much less time considering the implementation of those policies or the effects they had on teacher education organizations and the persons inside those places. Weick (1995) describes sense-making as the reciprocal interaction of information seeking, meaning ascription, and actions and claims that environmental scanning, interpretation, and associated responses provide. My colleagues and I were using an assortment of tools (most were fairly unsophisticated by today's standards) to understand the motives and actions of a range of policymakers across the political spectrum and to then attribute intent to their efforts.

The chapters in this book look to environments in which people now find themselves and attempt to explain how they got there. The chapters point to actions and events that have shaped the environments in which teacher education occurs. They highlight changing conditions and raise important questions about the intent of the actions. They point to new directions and the compelling need to press for new ways of preparing teachers for the nation's schools. The challenge of increased accountability for collegiate-based preparation programs and greater competition between those programs and alternative route programs is the reality that is described.

What seems unrecognized in the chapters is that the accountability and competition are the dominant themes for the foreseeable future. The people who enact school laws have reached the conclusion that there is little, if any, difference between beginning teachers prepared by colleges and universities and those who come to teaching through alternative routes (Gordon, Kane, & Stager, 2006). Using accountability measures that rely on score gains of students on standardized tests, policymakers have concluded that alternative route and traditional route teachers have about the same effect on student learning. As Hess notes in his commentary and elsewhere (2001), until there are other measures of student engagement and student success that are as acceptable or as compelling as the measures we currently use, student performance is *the* measure that matters.

We are confronted by an ambitious agenda from both ends of the political spectrum. As the focus for making teacher education policy has shifted from the states to the federal government (Ramirez, 2004), so too has the array of policies used to prompt important changes in teacher education:

- *Support for multiple pathways to teaching*—since the 1980s there has been a shift in the source for new teachers—from solely or primarily collegiate-based programs, largely regulated by the state, that produced the majority of beginning teachers, to a system of multiple providers that includes two-year colleges, school systems and regional or intermediate service centers, for-profit businesses, not-for-profit centers and traditional colleges and universities that are loosely controlled by states but which respond to market forces and candidate interests. The multiplicity of providers is not likely to diminish in the coming decade, with increasing competition between and among all of the providers.
- *A commitment to evidence*—the most compelling development in teacher education is the increased emphasis on evidence-based teacher education. Prompted by the efforts of the two professional accrediting agencies (TEAC and NCATE), the federal government's Title II reporting requirements, the commitment to longitudinal data collection, and the efforts of the Carnegie Corporation's Teachers for a New Era initiative, there is an enormous investment being made in data gathering and analysis. With passage of the 1998 amendments to the Higher Education Act, the federal government

challenged collegiate-based preparation programs to follow their graduates well beyond the point of graduation and to know their successes in K–12 schools. This book documents the surge of interest in making teacher education an evidence-based professional preparation program.

- *An insistence on partnerships*—faced with the reality that policymakers would no longer support college-based preparation programs, those seeking to influence policy a dozen years ago promoted the idea of college-school partnerships as a viable alternative. Partnerships were to benefit both the college and the school by serving as sites for student teaching and other practicum experiences, for experimentation and research, and often for the professional development of practicing teachers. Based on the descriptions of events in several states in this book and elsewhere, it is plausible to suggest that the insistence on partnerships has become a permanent feature of teacher education.

- *A skepticism about teacher professionalism*—with the advent of the *National Board for Professional Teaching Standards* (NBPTS), investment in autonomous professional standards boards, and the development of a more robust professional accreditation system in the 1980s, teaching may have come as close to realizing a profession of teaching as it will ever come. The description of the Indiana Professional Standards Board and its rise and fall as an independent autonomous body is one indication of how short-lived was this effort. Similarly, more than half of the education schools in the U.S. pursued professional accreditation and there were efforts made to require all preparation programs to be accredited. Since then, the system of professional controls has been largely undermined (as we saw in both Indiana and Florida) as alternatives have been created to both NBPTS and the National Council for Accreditation of Teacher Education and a reimagining of teaching as an occupation rather than a profession has taken hold.

- *The transformation of teaching*—the impact of *Teach for America* has been profound relative to the way that we see teaching in America's schools. Once seen as a career-long occupation, *Teach for America* has turned this concept on its head and legitimized the idea of short-termers making a difference. During the past decade teaching has come to be recognized as a field where there are both explorers and persisters—i.e., those who view teaching as a short-term public service commitment *vis a vis* those who see teaching as career long practice (phrases used by recent TFA teachers in doctoral courses I have been teaching). The explorers drop out after the first two or three years while the persisters spend their working years devoted to the variety of roles that constitute a professional career in teaching. This has implications for the way that teachers are prepared and schools are staffed.

- *Engagement of the arts and sciences*—a persistent theme in policymaking is the need for connection between teacher preparation and the arts and sciences. During the past decade there have been renewed efforts for beginning teachers to be more grounded in academic content with particular attention to the sciences and mathematics (so-called STEM initiatives), resulting in the restructuring of course requirements for secondary teacher candidates, and the redesign of elementary teacher preparation programs. The Common Core Standards Project will demand even greater alignment and cooperation between teacher education programs and their arts and science counterparts.

• *Changes in professional development for teachers*—one of the most profound changes with implications for teacher education (and, in particular, graduate education) is the growth of school-based professional development for teachers. Recent federal legislation has enabled schools and school districts to purchase and provide professional development for teachers from a host of providers—many for-profit. A parallel trend has been the rise in the school-based teacher development experiences that have become a substitute for traditional graduate courses and advanced degrees. Using new frameworks for professional development (e.g., communities of practice, teacher collaboratives, and situated learning), school- and district-sponsored professional development programs are drawing prospective graduate students away from formal university-based courses and programs.

• *A focus on children*—the rise in attention to special needs children and those from low-income backgrounds, often speaking a language other than English has caused a massive shift in the attention these children receive in preparation programs. Embracing these new responsibilities for schooling all children has created unprecedented demands on teacher education programs.

These developments or factors have come to overwhelm the vast majority of teacher education programs. None of the eight conditions briefly described above is likely to diminish or change. The need for more highly qualified teachers to teach more challenging children using the metrics of standardized achievement tests to measure both their own effectiveness and the gains of their students is likely to grow in prominence. The challenges to *professionalism* are likely to increase and the insistence on partnerships—across campus and with local schools—is likely to grow. The reality is that there will be more demands on the scarce resources available for teacher education and faculties, and deans will have to make even more strategic choices. Doing so in a time of intense competition and increased accountability makes the future most uncertain.

More than 40 years ago, B. Othanel Smith (1969) held up the possibilities of *reform vis a vis revolution* in teacher education. He said that these were the two pathways for those who sought to change teacher education—one was the reform of the existing programs and the other a radical restructuring of the enterprise that he labeled a form of revolution. After much consideration, Smith chose to pursue an agenda of *reform*. A host of prominent reformers followed in the wake of Bunny Smith's call for the reform of teacher education and offered dozens of reform strategies—longer or extended programs with more *life-space*, better integrated undergraduate programs, more college investment and involvement (with appeals to the all-university responsibilities), more theoretical or social foundations, more scholarship money to support better candidates, more time for candidates to spend in schools or other settings that serve children and youth, greater reliance on technology, more evidence gathering and documentation of progress by candidates (including work sampling), more emphasis on content knowledge or subject-matter majors in the subjects candidates will teach, vesting control over teacher education to other departments and programs, making teacher education a post-baccalaureate program, giving more attention to the language and cultural diversity of students, and making more explicit the commitment to the wellbeing of all children. Dozens of other strategies have been offered in hundreds of reports and thousands of articles that appeared in leading journals—all offering superb analyses of the problem of teacher education, but few posing solutions for the future.

Creating a Phoenix from the Ashes

In the set of cases and chapters included in this book, these are the conditions that face teacher education programs. This is a dreary set of challenges that collegiate-based teacher education faces. These challenges can be overwhelming to all but the hardiest teacher educator, but the need for teachers to teach children and youth will not go away. The Cochran-Smith and Fries' chapter (Chapter 9) frames an agenda of hope in an otherwise bleak landscape. We can imagine all sorts of substitutes for teachers in schools and classrooms in the future, but the reality is that we will need teachers to oversee the learning of all children. Achieving a greater sophistication of the technologies we use to teach and for students to learn, to assess that learning in new and more authentic ways, and to hold adults responsible for educating all children are shared and common goals across a vast spectrum of the policy and teacher education world. We may talk of warring camps and competing agendas, but the only agenda that counts is the one that leads to learning by all students. These realities invite thoughtful consideration as we seek to equip a new generation of learners to be the teachers of tomorrow.

References

Gordon, R., Kane, T., & Stager, D.O. (2006). *Identifying effective teaching using performance on the job*. Washington, DC: The Brookings Institution.

Hess, F.M. (2001). *Tear down this wall: The case for a radical overhaul of teacher certification*. Washington, DC: Progressive Policy Institute.

Ramirez, H.A. (2004). The shift from hands-off: The federal role in supporting and defining teacher quality. In F.M. Hess, A.J. Rotherham, & K. Walsh (Eds.), *A qualified teacher in every classroom?* Cambridge, MA: Harvard Education Press.

Smith, B.O. (1969). *Teachers for the real world*. Washington, DC: American Association of Colleges for Teacher Education.

Weick, K.E. (1995). *Sensemaking in organizations*. Thousand Oaks, CA: Sage Publications, Inc.

A MODEST APPROACH TO DECONSTRUCTING TEACHER EDUCATION POLICY

Penelope M. Earley

From the chapters presented in this book, in particular the three longitudinal policy case studies of events in Louisiana, Florida, and Indiana (Chapters 3, 4, and 6), we find a picture of a muddled teacher education policy system. It is a system under stress in which educators and policymakers claim to be working toward the same goals but are at odds with one another over how to get there.

In Chapter 2, essay author Frederick Hess suggests that the promotion of a value-added evaluation system to measure teacher effectiveness, as promoted in federal grant applications, is misguided and more measured strategies would be more appropriate. He cautions educators to stop being defensive about criticisms of the teacher education enterprise but does not acknowledge the demoralizing effect of shifting expectations on those who prepare teachers. In that same chapter, Neal McCluskey reviews Constitutional intent from the perspective of an *originalist* and wonders about the appropriateness of any substantial federal role in education.

We proposed using two policy frameworks to aid our understanding of teacher education policy issues: Deborah Stone's policy paradox and Datnow and Park's theory of sense-making co-construction (see Chapter 1). These frameworks are helpful to interpret the events around enactment of certain policies and the dynamics of attempting to interpret them. Attending to social justice as detailed in Cochran-Smith and Fries' chapter (Chapter 9) demonstrates Stone's point about policy being the clash of ideas and values. The attempts of education school deans to implement ever-changing policies in Florida present an example to apply sense-making co-construction theory. However, these helpful theories cannot explain how or why education policy overall and teacher education policy specifically has ended up in such a confused and contentious place.

Drawing upon the material presented in the first nine chapters, I offer a modest proposal to situate and deconstruct teacher education's relationship with policy mandates. Teacher education, of course, is not independent of, nor separate from, the larger K-16 enterprise so I begin with a discussion of the larger enterprise and then turn specifically to teacher education.

Peering Through a Different Looking Glass

What is it We Want?

The goal of education policy is to create an education system that meets the needs of everyone: children, parents, educators, people who do not have children, employers (civilian and military), and all elected and appointed decision makers. It is assumed that in this perfect environment satisfying everyone will require a system individualized to meet the needs and wants of all, but at the same time standardized so the success of each may be evaluated. It will occur in a variety of schools: public, independent secular, independent non-secular, public and non-public charters, online and face to face, and in homes with instruction provided by parents. All of these entities must be assured of sufficient resources to provide instruction, so there will be no bickering about public funding for one reducing funding for another. Through a carefully constructed ratio that defies mathematical laws, local, state, and federal governments will each have 51% control of these schooling entities. Families who object to a strong government role in education will have the option to send their children to unregulated schools. Students in K–12 schools will pursue a personalized yet standardized curriculum that stresses critical thinking skills, education basics, as well as all other topics of interest to students (think of these as mandatory electives). Children and youth will learn to reject intolerance through exposure to the ideas and values of a variety of cultures, except of course those values and cultures antithetical to those of their parents or caregivers. Because schooling will be so interesting and compelling there will be no dropouts, and all students will complete their K–12 education prepared to enter college, immediately join the workforce with the knowledge and skills needed by employers, or participate in volunteer activities for the betterment of society and the planet.

In this ideal environment, teachers and administrators will be selected only from the top 10% of college graduates, creating a slight deficit in medical school enrollments. These education professionals will receive preparation in a variety of settings approved by a quasi-governmental-semi-voluntary-kind-of-standardized-accrediting agency or agencies. Requirements for a teaching or administrative license will involve a range of assessments, written exams as well as long-term observations in a variety of educational settings. After six or seven years of these evaluations, they will be granted a teaching or administrator license. Salaries will be commensurate with the hourly rate of heart or brain surgeons (after all, teaching is a matter of the heart as well as the mind).

How do We Get What We Want?

To consider the challenge of how to reach this education pinnacle—a flawless education system that serves all, is best for all, and is agreed upon by all—I employ the metaphor of "a journey". For at least the last two decades, federal and state decision makers have been seeking the path that will lead them to an educational ideal. There have been recommendations on how to construct and organize school buildings, such as open classrooms. Scores of curricular alternatives have been offered, most recently expressed in recommendations from a national panel on mathematics. There have been suggestions regarding the length of the school year, optimum class size, and whether to make everyone who enters buildings walk through metal detectors. All of these ideas represent potential paths to the goal of a perfect education system but for this discussion it is not the merits or limitations of these paths; rather, it is the journey that is of interest.

Policymakers intending to reform public schools often form coalitions, traveling toward the education ideal in a bus; whereas those supporting private charter schools might be in a van. Others, such as advocates for a specific evaluation metric, used automobiles. Investors who expected a return on privately run schools typically were found in limousines. There are no roads to the education peak and no accurate maps to guide the way, but those in each vehicle believed they found a small path that will lead them to the pinnacle. Together and separately they raced off certain they will soon reach the apex. However, their journeys have not been successful. Some promising paths led to jungles covered with tangled vines. Other paths ended at swiftly flowing rivers that could not be traversed without a bridge, and a few were circular, with the policy reformers ending up exactly where they began. That portion of the education community supportive of online schooling remained back with their computers creating models to replace face-to-face instruction once the perfect education system is found. It is unfortunate that, as the vehicles sped off, a few ran over the hopes and dreams of some watchful children and youths.

After the various policy actors rushed off to find education nirvana, teacher educators gathered to decide what to do. Some argued they should wait for policymakers to pave the way and to beckon them to the new education world. From time to time a brave educator would suggest not following the policy vehicles at all but instead clearing their own path to the education ideal. When this happened, one or more senior teacher educators—scarred from years of policy battles—would remind everyone of State Rule 9.5336.28(a)(i), that all education vehicles must, by law, be equipped with a device that allows them only to drive on paths created by policymakers. Ultimately, the teacher educators decided the safest plan was to select a policy vehicle and follow it. Before leaving, a number of them invited school superintendents and other K-12 educators to join them as traveling partners. They even offered to share some federal grant money as an incentive. Only a few of the K-12 educators agreed. Most worried that forming new partnerships would take time and resources they needed to devote to test preparation for their students. Some quietly whispered that, if they were too closely aligned with teacher educators, policymakers may challenge the superintendents' commitment to reform. The teacher educators sped to catch up with the policymakers but, alas, many did so while keeping their eyes firmly focused in the rear-view mirror and consequently a number of crashes and clashes occurred when the teacher education cars collided with the policy makers' vehicles.

An observer might ask: Why didn't decision makers employ surveyors to study the terrain and advise everyone on the best path? Why weren't engineers asked to design and build roads and bridges for policymakers and those who followed? Were there no cartographers to employ GPS technology to prepare maps? These are excellent questions.

Reality

My parable may be extreme, but that does not mean it inaccurately represents actions of the policymaking and teacher education communities. During the second Bush administration, the path to educational excellence was based on an accountability system modeled, in part, on one used in Houston, Texas. This became central to the No Child Left Behind law. The Obama administration turned to reforms enacted in the Chicago Public Schools to model their education blueprint, which was put into place through the Race to the Top grant program. As Zeichner points out in Chapter 5, few reform initiatives are supported by strong empirical evidence, and major parts of the Bush and Obama agendas are no exception. Reform ideas often are put in place and then, after the fact, advocates look for supportive think tanks to evaluate them and "prove" they work. Would anyone willingly undergo open-heart surgery without some assurance that the medical community had already conducted the research to provide assurances that the procedure would have a good outcome? Enacting education reforms without evidence that they will help rather than harm children is tantamount to conducting unregulated experiments on them.

By the same token, teacher educators are repeatedly challenged to adjust programs to prepare teachers quickly and to have more sophisticated knowledge of their subject matter, child development, testing and measurement, use of data, and cultural sensitivity. In some cases teacher educators cling to the past, rationalizing that what they have been doing has an evidentiary base of sorts, so why change? Even if one acknowledges the claim of critics that some aspects of teacher preparation are constructed on a weak knowledge base, this does not justify abolishing them and jumping in another direction where there is no evidence at all of effectiveness. Researchers express frustration because determining policy effects may involve providing instructional interventions to one group of children but not another group; a strategy not acceptable to school personnel, parents, and university institutional research boards. Yet, by enacting untested policies, every day in every state children are subjects of policy experiments. Hess (Chapter 2) presents a sad truth: The policy world's distrust of the current education system causes them to discount empirical evidence presented by education researchers and give disproportionate weight to scholarship conducted by individuals outside the education system.

Realistically there can never be an education system that perfectly meets the needs and wants of every child, every parent, every potential employer, and every decision maker. That does not mean we cannot do better. Perhaps it is time to impose a moratorium on any new federal or state education policy initiatives. Charge each governor with engaging in a year-long series of conversations with constituents, not about how to change schooling, but about an agreed-upon purpose. These conversations should consider what schools can and should do to support the purpose and what will require partnerships with social service agencies, community groups, and parents. These discussions must also address the contentious issue of what level or levels of government should have authority over which education policy decisions. After that, researchers, under the umbrella of the National Academy of Education, should identify existing evidence to inform how to achieve the agreed-upon purpose and gaps in the knowledge base. Once these fundamental steps are taken, the system of teacher preparation can be reviewed. It may be that the way in which education professionals (teachers and administrators) are prepared needs to be revised, perhaps it needs to be replaced with something entirely different, or the current system may be best for the task. I will not speculate on the future of teacher education, but do recognize that educators are frustrated with policy whims that continually expect them to adjust their professional practice and that decision makers are frustrated when change is not immediate.

A Final Word

The chapters prepared for this book are not merely theoretical or academic exercises. The authors are parents, aunts, uncles, and grandparents. They may disagree on the role of government, the ideal way to structure schools, and priorities for teacher preparation, but each and every one cares about all children and the future that awaits them.

GLOSSARY

AACTE: The Washington, DC-based American Association of Colleges for Teacher Education is a national association of schools, colleges and departments of education. The association has 800 member institutions in the states and territories. In some states there are affiliated chapters, such as the Indiana Association of Colleges for Teacher Education (IACTE) or the Florida Association of Colleges for Teacher Education (FACTE). For more information, go to www.aacte.org

ABCTE: The American Board for Certification of Teacher Excellence is a Washington, DC-based alternative certification organization. Established with monies from the US Department of Education, this 501(c)(3) organization recruits and prepares career changers for a test of teaching skills. ABCTE's certification is accepted in 10 states. For more information, go to www.abcte.org

Achieve: A Washington, DC interest group that was formed by the nation's governors and business leaders in 1996. The group helps states raise academic standards and graduation requirements, improves the assessments used and strengthens accountability. For more information, go to www.achieve.org

AERA: Founded in 1916, this Washington, DC-based organization of 25,000 academics and researchers advocates for scholarly inquiry related to education and evaluation and promotes the practical application of research results. For more information go to: www.aera.org.

AFT: The American Federation of Teachers, established in 1916, is a union of professionals that is located in Washington, DC. Its some 1.5 million members are primarily located in urban school districts, the health professions, local and state government, and on college campuses. It promotes collective bargaining for teachers and other educational personnel, with the goal of improving the lives of its members and their families. For more information, go to www.aft.org

Alternative certification (also **Alternative routes**): Alternative routes to teacher certification are state-defined routes through which individuals with at least a baccalaureate degree can obtain licensure. These programs are offered by two- and four-year colleges and universities, not-for-profit and for-profit entities, local education agencies, and others. In 2010, 48 states and the District of Columbia had approximately 600 alternative route programs that produced some 60,000 beginning teachers. In some cases, candidates may be employed as educators while earning a teaching license.

ARRA: The American Recovery and Reinvestment Act of 2009 (PL 111–5) is the economic stimulus package enacted in February of that year and signed into law by President Barack Obama. Included in ARRA is $100 billion in monies for education programs.

CCSSO: The Council of Chief State School Officers, is a non-partisan nation-wide non-profit organization that is based in Washington, DC for public officials who head departments of education at the state level. CCSSO is a state-led organization that leads and facilitates collective state action to enhance education in the states. For more information, go to www.ccsso.org

Clinical Faculty: School and higher education personnel responsible for instruction, supervision, and assessment of teacher candidates during their field experiences and clinical practice.

Common Core Standards Initiative: An initiative of the NGA and CCSSO that is seeking to create a set of state led education standards in English language arts and mathematics for Grades 1–12. They are being developed in collaboration with a variety of stakeholders that include subject-matter experts, teachers, school leaders, and parents. For more information, go to www.ccsso.org

DOE (see **SEA**)

ESEA (also **NCLB**): The Elementary and Secondary Education Act of 1965, signed into law by President Lyndon B. Johnson as part of the War on Poverty, extended the role of the federal government in elementary and secondary education. ESEA emphasized equal access to education and a major purpose of the Act was to provide funds for the education of poor children in so-called Title I schools. The law authorized a range of programs that are administered by the States. In 2001, the law was reauthorized with the name No Child Left Behind Act and included greater emphasis on high standards and accountability for results. Title II of the current law focuses on teacher preparation and the recruitment of highly qualified teachers and principals.

ETS: The Princeton, NJ-based Education Testing Service is a non-profit organization with the mission of advancing quality and equity in education for people worldwide. It administers and scores 50 million tests annually in more than 180 countries. Among those tests are ones used to assess the knowledge, skills and competencies of beginning teachers. The so-called Praxis Series is administered to approximately 125,000 teacher candidates each year as part of the certification process required in some states (some states use different exams). For more information, go to www.ets.org

GAO: The General Accountability Office is the investigative arm of Congress charged with auditing and evaluating government programs and activities. It is an independent and non-partisan agency that supports the Congress in meeting its constitutional responsibilities and to ensure the accountability of the federal government to the American people.

HEA: The Higher Education Act of 1965 was signed into law by President Lyndon B. Johnson in November 1965. PL 89–329 was part of the Great Society legislation and was intended to strengthen the educational resources of the nation's colleges and universities and to provide financial assistance to students in postsecondary education.

Holmes Partnership (Holmes Group): The Holmes Partnership is a consortium of 28 universities, public school districts, teacher associations, and local as well as national organizations currently based at the University of Florida in Gainesville. Its focus is on partnerships to serve as a major vehicle for the reform of teaching and learning in K–12 schools and colleges and universities. The group was formed in 1995 from the earlier Holmes Group, named in honor of Harvard Graduate School of Education's inaugural dean, Henry Wyman Holmes. For more information, go to www.holmespartnership.org

INTASC: The Interstate New Teacher Assessment and Support Consortium, a project of the Council of Chief State School Officers (CCSSOs) that developed model performance-based standards and assessments for the licensure of teachers.

LEA: Local Education Agencies are school districts or school divisions (VA) that support and operate elementary, middle, and secondary schools in the U.S. They have powers similar to that of a county or city. There were 17,735 operating LEAs in the U.S. in 2008–2009.

Licensure: The official recognition by a state governmental agency or professional standards board that an individual has met certain qualifications specified by the state and is, therefore, approved to teach in the public schools of the state. In some cases the license is referred to as a teaching certificate.

NBPTS: The National Board for Professional Teaching Standards, an organization of teachers and other educators, formed in 1987 to advance quality education and learning, has developed standards and a system for assessing the performance of experienced teachers seeking national certification. Some 82,000 teachers are now National Board Certified. For more information, go to www.nbpts.org

NCATE: The National Council for Accreditation of Teacher Education is one of two national accrediting bodies in education recognized by the United States Department of Education (the other is the Teacher Education Accreditation Council or TEAC). Talks are underway for a merger of NCATE and TEAC. Accreditation is defined as a process for assessing and enhancing academic and educational quality through peer review. Some states require accreditation for its colleges and universities offering teacher education; others see it as voluntary. In 2010, 657 colleges were accredited by NCATE. NCATE review determines whether or not the unit responsible for teacher education at a given college or university meets its standards for approval. Accreditation is for seven years and is renewed only after review by a team of educators. For more information, go to www.ncate.org

NCLB: No Child Left Behind is the name given to the version of the Elementary and Secondary Education Act (ESEA) passed by Congress and signed by President George W. Bush in 2001. It differs from earlier versions of ESEA in several important ways. First, it offers a definition of a "highly qualified teacher," the first time the federal government has offered such a definition. Second it requires "adequate yearly progress" of schools with the consequence of closing schools not meeting the standards for progress. Third it requires the disaggregation of data reported so that differences among races can be seen. It has been criticized by some for its reliance on standardized testing to determine progress and academic success. For more information go to www.ed.gov//nclb/landing.jhtml

NCTAF: The National Commission for Teaching and America's Future was founded in 1994 at Teachers College (Columbia University) and became a non-profit research advocacy organization based in Washington in 2002. NCTAF engages in research and writes policy papers related to quality teaching, including teacher education, teacher assessment, teacher attrition, induction into teaching, school segregation, and professional development. NCTAF has developed partnerships with 24 states to advance its work and establish demonstration sites. For more information, go to www.nctaf.org

NCTQ: The National Center for Teacher Quality is an independent advocacy and research group that studies the effectiveness of states in providing what they define as well prepared teachers, an adequate pool of teachers, success in identifying effective teachers, retaining them, and removing ineffective teachers. The organization is a strong advocate of alternate routes to certification and criticizes most states for requiring too many professional courses of future teachers. For more information, go to www.nctq.org

NEA: The National Education Association is the largest teacher union in the United States. The NEA traces its roots to 1857 and now represents 2.7 million members. The union takes positions on issues beyond teacher pay, including policy issues such as the evaluation of teachers and federal support for education. For more information, go to www.nea.org

NES: National Evaluation Systems (now Pearson Evaluation Systems Group) is a testing company begun in Massachusetts in 1972 that provides teacher certification tests to a number of states, including Arizona, California, Colorado, Georgia, Florida, Illinois, Massachusetts, Michigan, New Mexico, New York, Oklahoma, Oregon, Texas, Virginia, and Washington. NES tests are customized to reflect state standards and available to test basic skills, content, and pedagogical knowledge. For more information, go to http://education.pearsonassessments.com/pai/ea/teacher/teacherhome.htm

NGA: The National Governor's Association is a non-partisan association of the nation's governors. The association holds annual meetings with a focus on policy issues and publishes research and summaries of best practices in various policy areas, including education. For more information, go to www.nga.org

NNER: The National Network for Educational Renewal is a network of partnerships between colleges and universities offering teacher education programs and K–12 schools. The NNER was founded by John I. Goodlad and is now an independent organization advocating for the renewal of teacher education and P–12 schools, with a focus on the purposes of education in a democracy. In each setting the network promotes collaboration among faculty in education, in arts and science, and in schools. More than 40 college/university and school partnerships constitute the network. For more information, go to www.nnerpartnerships.org

Normal School: "Normal School" was the name given to teacher education institutions, primarily those preparing elementary teachers. The name is derived from one of the first such schools, Ecole Normale Supérieure (or Normal Superior School), established in Paris in 1794. Normal Schools were the primary institutions preparing teachers in the United States through the early 20th Century. Many Normal Schools have evolved into state colleges and universities.

NRC: The National Research Council (NRC) functions under the auspices of the National Academy of Sciences (NAS), the National Academy of Engineering (NAE), and the Institute of Medicine (IOM). The Council was established by Congress during the administration of Abraham Lincoln, but operates as a private, not-for-profit organization. Within the NRC there is a division of behavioral sciences, social sciences, and education, which issues reports on education that are often influential. The Council promotes work that is scientifically based, studies the use of tests, reading, learning, and the preparation of teachers. An example of a recent report is *Preparing Teachers: Building Evidence for Sound Policy*. For more information, go to http://sites.nationalacademies.org/NRC/index.htm

PACT: In response to fall 1998 legislation in California, SB 2042, a testing system known as Performance Assessment for California Teachers, has been developed to assess the likelihood of success of beginning teachers and is widely used in California. Since then the test has had wide use and, through the American Association of Colleges for Teacher Education and the Council of Chief State School Officers (the national organization of state commissioners and secretaries of education), an adaptation of PACT, known as the Teacher Performance Assessment, is being tested in a number of states. Field tests are underway in a number of states. For more information on PACT, go to www.pacttpa.org/_main/hub.php?pageName=Home. For information on the AACTE/CCSCO project, go to http://aacte.org/index.php?/Programs/Teacher-Performance-Assessment-Consortium-TPAC/teacher-performance-assessment-consortium.html

PDS: A Professional Development School, a concept originally developed by the Holmes Group, is a school in partnership with a college or university designed to prepare teachers and enhance school practices. It has been likened to a teaching hospital, as a site where future teachers study in a real-world context. Many colleges and universities have developed Professional Development Schools, although they differ greatly in their structure and goals.

Professional Standards and Practices Board: A Professional Standards and Practices Board is an entity established by a state to either make or advise on policies for the certification of teachers and the standards for teachers. Most are advisory to state boards of education that have the power to make policy, but some are independent and are empowered to make policy.

PTA: The Parent Teacher Association was originally known as the National Congress of Mothers and from its inception it was an advocacy group for quality education for children. Many local schools have PTAs as a means of involving parents in schoolwork. There are state-level PTAs and a National PTA in Washington which takes positions on major policy issues. For more information, go to www.pta.org

RTTT (also **R2T** and **RTT**): Race to the Top is a $4.5 billion U.S. Education Department program designed to spur reforms in states and local school districts. The focus of the program is to improve teacher and principal effectiveness based on performance. The program supports high quality pathways for aspiring teachers and principals, effective support for teachers, and improvements in the effectiveness of teacher and principal preparation programs.

SEA: State Education Agencies, or State Departments of Education, are found in all 50 states and are government agencies responsible for carrying out state-wide policy and monitoring of education, often with funding of public schools, and the certification of teachers and monitoring teacher quality. They are typically the agencies that carry out policy of the State Board of Education and are headed by a Commissioner of Education, a Superintendent of Public Instruction, or an executive holding a similar title.

State Board of Education: The State Board of Education, sometimes known by a somewhat different name (i.e., the Texas Education Agency), is the appointed or elected body charged with making state policy in education. Some states have a separate board for higher education. Some, like the New York Board of Regents, are responsible for P-12 education, the certification of teachers, higher education, museums, libraries, and the licensing of other professionals.

State Program Approval: In many states the process of authorizing a college or university to offer a teacher education program involves state program approval, the approval by the state of a specific program for preparing teachers. Program approval is often renewed periodically, or some states defer to national accreditation for approval. States using program approval accept the recommendation of the college or university as the basis for issuing a license to teach.

TEAC: The Teacher Education Accrediting Council is one of two national accrediting bodies in education recognized by the United States Department of Education (the other is the National Council for Teacher Accreditation or NCATE). TEAC was begun as an alternative to NCATE in 1997 and differs from NCATE in that accreditation is built around the program's case that is, in turn, built by the institution demonstrating with evidence that it prepares competent, caring, and qualified professional educators. Unlike NCATE, there is not a set of specific standards to be met. NCATE and TEAC are in talks to secure a merger of the two agencies. For more information, go to www.teac.org

Teaching license (Teaching certificate): A teaching license is a credential issued by a state allowing one to teach in a public school after completing state requirements for certification. Requirements usually include coursework, clinical experiences, and passing tests in content and teaching practices. Certificates apply to particular grade levels and subjects, such as elementary school, high school mathematics, or middle school science.

TNE: Teachers for a New Era is a grant program originated by the Carnegie Corporation of New York, and supported as well by the Ford Foundation and the Annenberg Foundation. Eleven teacher education institutions were funded to develop programs based on three design principles, including using evidence to improve programs that includes evidence of student learning, involvement of the arts and science faculty in teacher education, and defining teaching as an academically taught clinical practice profession. Funding for the project has now ended. For more information, go to www.teachersforanewera.org

Value-added assessment (also **Value-added model** and **Value-added teacher preparation assessment**): Largely credited to the work of William Sanders, then at the University of Tennessee, who developed a statistical model for tracking the effect of an individual teacher on the change in standardized test scores of students. Some states and school districts began using the model as a way of evaluating teachers. The system, which is controversial, because often it is focused on to the exclusion of other measures and because some people doubt its validity, has been elevated by the Race to the Top funding competition of the Obama administration. An absolute criterion for participation in Race to the Top was that there were no laws or regulations that precluded the use of student test scores in the evaluation of teachers.

CONTRIBUTORS

Frank Alvarez has been Superintendent of the Montclair Public Schools since 2003. Prior to that, he served as Superintendent of Schools in North Caldwell, NJ and as principal of two schools in Montclair. He serves on the boards of a number of education organizations, including the Minority Student Achievement Network, the Tri-State Consortium, and the Victoria Foundation.

Marilyn Cochran-Smith is the Cawthorne Professor of Teacher Education for Urban Schools and Director of the Doctoral Program in Curriculum and Instruction at the Lynch School of Education at Boston College. A former President of the American Educational Research Association, she is an elected member of the National Academy of Education.

Elayne Colón is the Director of Assessment and Accreditation in the College of Education at the University of Florida. Dr Colón has published journal articles in peer-refereed journals, including the *Journal of Psychoeducational Assessment* and *Journal of Special Education*, and presented numerous papers at national and state conferences.

Ada Beth Cutler is Professor and Dean of the College of Education and Human Services at Montclair State University. Previously she served as a K–8 principal and Senior Research Associate at Education Matters, and as a policy analyst at the Consortium for Policy Research at Rutgers University. She is the former director of Montclair State University Network for Educational Renewal.

Thomas Dana is Associate Dean for Academic Affairs in the College of Education at the University of Florida. He previously served as director of the School of

Teaching and Learning. From 1998 to 2003, he held the Henry J. Hermanowicz Professorship in Education and coordinated teacher education programs.

Van Dempsey is Professor and Dean of the School of Education and Human Performance at Fairmont State University in West Virginia. He has provided leadership at the local, state, and national levels in efforts to enhance teacher education, professional development, and school-university collaboration to improve public education.

Penelope M. Earley is Professor and Founding Director of the Center for Education Policy and Evaluation at George Mason University. She has published multiple articles and chapters on education policy and co-edits the *Journal of Education Policy and Evaluation*. Before moving to Mason, she was a Vice President at AACTE.

Catherine Emihovich is Professor and Dean of the College of Education at the University of Florida. She previously served as Dean of the College of Education at California State University-Sacramento, and has held academic positions at the State University of New York at Buffalo, Florida State University, and the University of South Carolina.

Patricia D. Exner has 35 years of professional experience in K-12 and higher education. Currently she is Associate Dean in the College of Education at Louisiana State University. A participant in Louisiana's Blue Ribbon Commission for Educational Excellence meetings since 1999, she has been actively involved in teacher education redesign and accountability reforms in Louisiana.

M. Jayne Fleener is currently the Dean of the College of Education at North Carolina State University and has over 25 years of professional experience in K-12 and higher education, including teaching high school mathematics and computer science in North Carolina. Previously, she was the Dean of the College of Education at Louisiana State University, where she also held the E.B. ("Ted") Robert professorship.

Kim Fries is an Associate Professor of Education at the University of New Hampshire. She is involved in state policy issues regarding teacher preparation and quality as well as university and K-12 school policy. Her research interests include teacher education, the policies and politics surrounding teacher quality, and teacher education.

Frederick Hess is Resident Scholar and Director of Education Policy Studies at the American Enterprise Institute. An educator, political scientist, and author,

Hess studies a range of K-12 and higher education issues and is the author of many influential books on education.

David G. Imig is a Professor of Practice at the University of Maryland, College Park, and directs the Carnegie Foundation for the Advancement of Teaching Project on the Education Doctorate, a consortium of 24 colleges and universities invested in transforming their professional practice doctorate. He is AACTE President Emeritus.

Neal McCluskey is the Associate Director of Cato's Center for Educational Freedom. Prior to arriving at Cato, McCluskey served in the U.S. Army, taught high school English, and was a freelance reporter covering municipal government and education in suburban New Jersey.

Nicholas M. Michelli is Presidential Professor in the Ph.D. Program in Urban Education at the Graduate Center of the City University of New York. He recently served as visiting professor in Shanghai and Hong Kong. Prior to that he was a dean for 25 years. He writes, speaks, and consults about educational policy, urban education, teacher education, democracy and social justice in education, and large-scale assessments.

Deborah Shanley is Professor and Dean of the School of Education at Brooklyn College in Brooklyn, New York. Her current work focuses on creating collaborations and partnerships with New York City schools and cultural institutions as vehicles for extending teacher preparation beyond the walls of academia. She chairs the Governing Council of the National Network for Educational Renewal.

Susan Taylor recently retired after 37 years as an educator and administrator in the Newark Public Schools. She is currently the Director of the Newark-Montclair Urban Teacher Residency Program funded by a $6.3 million Teacher Quality Partnership Grant from the U.S. Department of Education.

Theresa Vernetson is the Assistant Dean for Student Affairs in the College of Education at the University of Florida. She serves as the College's liaison with the Florida Department of Education.

E. Lynne Weisenbach is Vice Chancellor at the Board of Regents of the University System of Georgia. Prior to coming to Georgia, Weisenbach served as Dean of the College of Education and as a Professor at the University of Indianapolis from 1993 to 2008. In 2001, She founded the Center of Excellence in Leadership of Learning (CELL).

Ken Zeichner is the Boeing Professor of Teacher Education and the Director of Teacher Education at the University of Washington, Seattle. He is a member of the National Academy of Education, was co-chair of the Panel on Research in Teacher Education of the American Educational Research Association (AERA), and was Vice President of AERA's division Teaching and Teacher Education from 1996 to 1998.

INDEX

Taylor & Francis

eBooks

FOR LIBRARIES

ORDER YOUR FREE 30 DAY INSTITUTIONAL TRIAL TODAY!

Over 23,000 eBook titles in the Humanities, Social Sciences, STM and Law from some of the world's leading imprints.

Choose from a range of subject packages or create your own!

Benefits for **you**

▶ Free MARC records
▶ COUNTER-compliant usage statistics
▶ Flexible purchase and pricing options

Benefits for your **user**

▶ Off-site, anytime access via Athens or referring URL
▶ Print or copy pages or chapters
▶ Full content search
▶ Bookmark, highlight and annotate text
▶ Access to thousands of pages of quality research at the click of a button

For more information, pricing enquiries or to order a free trial, contact your local online sales team.

UK and Rest of World: **online.sales@tandf.co.uk**

US, Canada and Latin America:
e-reference@taylorandfrancis.com

www.ebooksubscriptions.com

ALPSP Award for BEST eBOOK PUBLISHER 2009 Finalist
sponsored by

Taylor & Francis **eBooks**
Taylor & Francis Group

A flexible and dynamic resource for teaching, learning and research.